# DIGITAL TECHNIQUES

*Macmillan Technician Series*

P. Astley, *Engineering Drawing and Design II*

P. J. Avard and J. Cross, *Workshop Processes and Materials I*

G. D. Bishop, *Electronics II*

G. D. Bishop, *Electronics III*

J. C. Cluley, *Electrical Drawing I*

J. Elliott, *Building Science and Materials*

John G. Ellis, and Norman J. Riches, *Safety and Laboratory Practice*

D. E. Hewitt, *Engineering Science II*

P. R. Lancaster and D. Mitchell, *Mechanical Science III*

R. Lewis, *Physical Science I*

Noel M. Morris, *Digital Techniques*

Noel M. Morris, *Electrical Principles II*

Noel M. Morris, *Electrical Principles III*

Owen Perry and Joyce Perry, *Mathematics I*

# DIGITAL TECHNIQUES

Noel M. Morris

*Principal Lecturer,*
*North Staffordshire Polytechnic*

M

*First published 1979 by*
THE MACMILLAN PRESS LTD
*London and Basingstoke*
*Associated companies in Delhi Dublin*
*Hong Kong Johannesburg Lagos Melbourne*
*New York Singapore and Tokyo*

*Typeset in 10/12 Times*
*Printed in Great Britain by A. Wheaton & Co. Ltd., Exeter*

# Contents

# Preface

The explosive growth of digital electronics has affected the life of almost everyone. This book provides coverage of aspects of digital technology ranging from basic gates to the design of digital systems.

Digital technology is wide-ranging and this fact is reflected in TEC Certificate and Diploma courses as well as in TEC Higher Certificate and Diploma courses. The aim of the book is to provide a sound introduction to Digital Techniques associated with these courses. Students attending other courses which include digital technology will also benefit from reading the book.

The eleven chapters provide a solid introduction to digital technology beginning with chapters on basic logic functions and gates, leading to the types of semiconductors used and to integrated circuit technology. A wide range of logic families are described including DTL, TTL, ECL, MOS and $I^2L$. There follows two chapters devoted to Boolean algebra and to the use of Karnaugh maps, which form the basis of many design studies. Binary arithmetic methods are introduced in chapter 9, together with binary codes and error detection. The ubiquitous pocket calculator and the microprocessor could not operate without counters, shift registers and storage systems and these are introduced in chapters 10 and 11. Included in the final chapter are semiconductor storage devices including RAMs and ROMs together with ferrite ring core stores, magnetic bubble devices and charge coupled devices.

I would like to acknowledge the assistance, tolerant understanding and encouragement of my wife during the preparation of this book.

NOEL M. MORRIS

# 1 The Basis of Logic Systems

## 1.1 GATES

The subject of electronic logic is one which embraces the whole field of electronics from computers to automobiles and from telephone exchanges to toys. 'Logic' devices serve man in every walk of life, each device or system operating in a predictable manner. In fact, so predictable is their operation that we can use a form of *logical algebra* to determine the way in which the circuit works. This type of algebra is sometimes known as *Boolean algebra*, after the Rev. G. Boole (1815–64) who set down the basic rules. Mathematicians often refer to it as *set theory*, or the theory of 'sets'.

The basic rules of this form of algebra are quite simple and, once understood, are relatively easy to apply. Devices used in logic networks control the flow of *information* through the system and for this reason are known as *logic gates*, since the 'gates' are opened and closed by the sequence of events occurring at their inputs. The basic range of gates are known by the names *AND*, *OR*, *NOT*, *NOR*, and *NAND*, and are described in detail in the chapters which follow.

The operation of each gate or system is defined by a logical algebraic statement, the logical equation being amenable to manipulation by the rules of Boolean algebra. Thus we see that Boolean algebra is a means of setting down the operation of logic circuits and networks in the form of a series of 'equations'.

Many applications of logic networks, such as those involved in counting and other arithmetic processes, require the use of *MEMORY* elements which retain or *store* information. Memory elements are constructed simply by interconnecting a number of the basic gates mentioned above in such a way that the circuit retains the original input data after the input signal has been removed. The information stored by the memory can be changed or updated by subsequent control signals.

## 1.2 LOGIC SIGNAL LEVELS

In the world of logical algebra every question has a definite solution, so that all problems give a 'yes' or 'no' type of solution, that is, the solution is either *true* or *false*. Thus, we are dealing with a *binary* or two-level system.

Digital electronic circuits have specific bands of voltage levels allocated to them which conform to the 'true' and 'false' conditions. If, for example, we have an electronic switch whose supply voltage is + 5 V then, depending on the external load connected, the 'true' output condition may be represented by any voltage in the range + 3 to + 5 V, and the 'false' output may be represented by a voltage in the range zero to + 0.5 V. Using a notation known as *positive logic*, the 3–5 V range is described as the logic '1' level, and the 0–0.5 V range is described as the logic '0' level. In this notation, logic '1' is a 'true' solution, and logic '0' is a 'false' solution.

Another notation which is sometimes used is the *negative logic* notation, in which the 3–5 V range is described as logic '0', and the 0–0.5 V range is described as logic '1'.

The positive logic notation is frequently used in connection with logic circuits, although there are exceptions to this, which are mentioned as they arise in the text.

## PROBLEMS

**1.1** A binary system is a two-level system. How many 'levels' exist in (a) ternary, (b) octal, (c) duodecimal and (d) hexadecimal systems?

**1.2** A binary logic gate has operational voltage levels of + 3.5 V and − 2 V. State which corresponds to logic '0' in (a) negative logic, (b) positive logic.

**1.3** Give four examples of two-state devices.

# 2 Basic Logic Functions

In order to outline the operation of the logic devices mentioned in chapter 1, let us consider the operation of a coin circuit of a hypothetical coin-operated vending machine.

Suppose that our machine dispenses a beverage after either a 10p or a 5p coin has been inserted into a coin slot. The cost of our drink is to be 5p, so that when a 10p coin is inserted, the machine returns 5p to us. If a 5p 'change' coin is not available in the machine, a sign stating 'USE CORRECT CHANGE ONLY' is to be illuminated. In the following sections we will consider how the functions of the coin circuit may be carried out by logic gates.

## 2.1 THE AND FUNCTION

In this section we shall consider the circuit associated with the release of the 5p change. Prior to inserting a 10p coin we must take note that the 'USE CORRECT CHANGE ONLY' light is extinguished, which implies that the machine holds at least one 5p coin. Thus, when the 10p coin is inserted AND the 'CORRECT CHANGE' light is extinguished, then the coin-release circuit is activated. If we assign symbol $T$ to the output of the detector which senses the 10p piece, symbol $C$ to the output of the sensor associated with the 5p coin stored in the machine, and symbol $X$ to the output of the logic gate which initiates the coin release mechanism, then

$$X = T \text{ AND } C = T . C$$

The 'dot' (.) symbol is used in logical equations to represent the logic AND function.

Sensor $T$ provides a logic count '1' output signal when a 10p coin is inserted, and a logic '0' when no 10p coin is present. Similarly, the output from sensor $C$ is '1' when the machine holds a 5p coin, and a '0' when it does not. Thus $X = 1$ (that is, a 5p coin is released) when $T = 1$ AND $C = 1$ simultaneously. If either $T = 0$ or $C = 0$, then $X = 0$ and the operation of the coin release circuit is inhibited.

If we collect together all the possible operating combinations of the AND gate associated with the coin-release circuit, we have what is known as a *truth table* which, for the two-input AND gate, is given in table 2.1.

**Table 2.1 Truth table for a two-input AND gate**

| Inputs | | Output |
|---|---|---|
| $T$ | $C$ | $X = T.C$ |
| 0 | 0 | 0 |
| 0 | 1 | 0 |
| 1 | 0 | 0 |
| 1 | 1 | 1 |

A relay circuit which satisfies the AND function is shown in figure 2.1. Here the signals $T$ and $C$ are derived from the sensors described above, and output $X$ is used to energise the coin-release circuit. In our relay circuit, when either signal $T$ or signal $C$ is logic '0', the appropriate relay contacts are open and the output voltage from the circuit is zero. When $T = C = 1$, both contacts are closed and the full voltage (logic '1') appears at the output. A number of symbols are used to represent the AND gate, two popular versions being shown in figure 2.1.

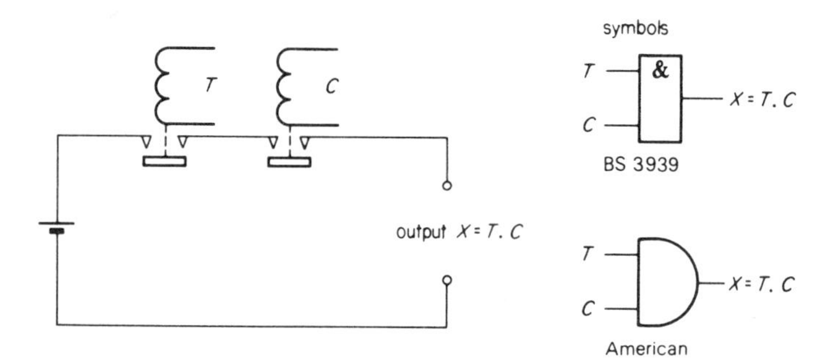


Figure 2.1 A logic AND gate

In a complex system the AND gate may have a large number of inputs, say $N$ inputs, and if the output from the circuit is $f$, then the logical equation describing the operation of the gate is

$$f = A.B.C. \ldots L.M.N$$

This corresponds to a relay circuit with $N$ relays in series, each with a normally open pair of contacts and each relay being energised by a separate input signal.

## 2.2 THE OR FUNCTION

In the specification of our vending machine we said that the drink was prepared if either a 10p coin or a 5p coin is inserted into the coin slot. If we assign the symbol $F$ to the sensor which detects 5p pieces and, as before, sensor $T$ detects 10p coins, and we let symbol $Y$ represent the output from the gate, then

$$Y = F \text{ OR } T = F + T$$

The 'plus' $(+)$ symbol is used throughout this book to represent the logical OR function. This symbol should not be confused with the arithmetic addition symbol, the difference between the two symbols being explained later in this section. An alternative symbol sometimes used for the OR function is a 'vee' (v), that is, $Y = F \vee T$. The truth table for the OR function described above is given in table 2.2.

**Table 2.2 Truth table for a two-input OR gate**

| Inputs | | Output |
|---|---|---|
| $F$ | $T$ | $Y = F + T$ |
| 0 | 0 | 0 |
| 0 | 1 | 1 |
| 1 | 0 | 1 |
| 1 | 1 | 1 |

From the truth table we see that the output from the gate is '1' when either input signal is '1'. An interesting situation arises if we insert a 5p and a 10p coin simultaneously (assuming, of course, that this can be done!), since the machine accepts both coins and dispenses only one drink! This condition is illustrated in the final line of the truth table, in which the output signal from the gate is '1' when both inputs are activated. Thus the logical statement that 1 + 1 = 1 is valid; it should *not* be confused with the arithmetic addition operation.

Also, from the truth table, the output from the gate is zero when both inputs are zero, that is, when no coins are inserted.

A logic '1' output from the OR gate causes the vending circuits to be activated, so producing our drink.

A relay circuit which satisfies the OR function is shown in figure 2.2. In this case the circuit provides an output voltage when either $F$ OR $T$ OR both input signals are present. This circuit is sometimes known as an *inclusive–OR* gate, since it provides an output signal in the case when $F$ and $T$ are both '1'. Later in the book we shall deal with yet another type of gate known as an *exclusive–OR* gate, which gives an output of logic '0' when both $F$ and $T$ are energised by logic '1' signals.

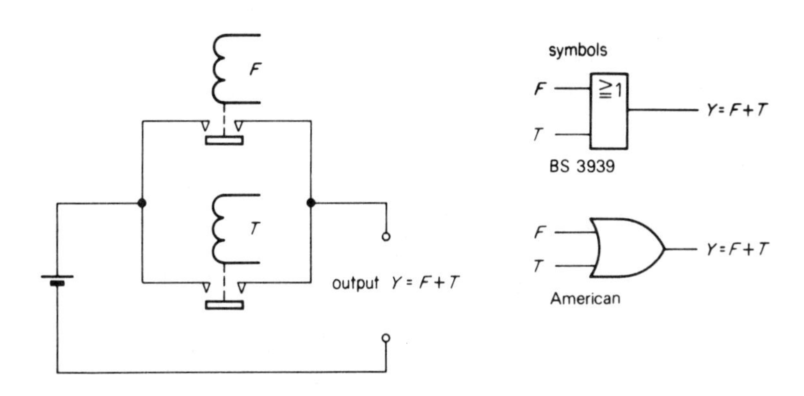


Figure 2.2 A logic OR gate

Some circuits require the use of multiple-input OR gates, and if a large number of inputs, say $N$ inputs, are employed then the logical output $f$ developed by the gate is

$$f = A + B + C + \ldots + L + M + N$$

This corresponds to a relay circuit with $N$ relays in parallel, each having a normally open pair of contacts and each relay being energised by a separate signal.

## 2.3 THE NOT FUNCTION

When discussing the signal levels associated with logic circuits, we saw that only two steady values could exist, that is, logic '1' and logic '0'.

When the output from a logic gate is '1', quite clearly it is NOT '0'. Also, when the output is '0' it is NOT '1'. Thus a gate which generates the NOT function provides an output of logic '0' when the input is '1' and vice versa. The process of logically *inverting* or *complementing* (that is, the NOT function) a function is signified by placing a 'bar' over the function, as shown below for the function $C$.

$$\text{NOT } C = \overline{C}$$

The truth table for this function is given in table 2.3.

**Table 2.3 Truth table for a NOT gate**

| *Input* $C$ | *Output* $\overline{C}$ |
|---|---|
| 0 | 1 |
| 1 | 0 |

Let us consider how a NOT gate can be used in the coin circuit of our vending machine. In the original specification of the machine we said that if a 5p piece is not available in the machine, then the 'USE CORRECT CHANGE ONLY' light is to be illuminated.

Hence, we can illuminate the lamp from the output of a NOT gate whose input is energised from sensor $C$. You will recall that sensor $C$ is used to detect the presence of coins held in the 'change' stack inside the machine. Thus when change is NOT held in the machine, the output from sensor $C$ is zero, so that $\overline{C} = 1$ and the change warning lamp is illuminated. When sensor $C$ detects a 5p piece in the machine, the output from the sensor $C$ is '1' so that $\overline{C} = 0$ and the warning lamp is extinguished.

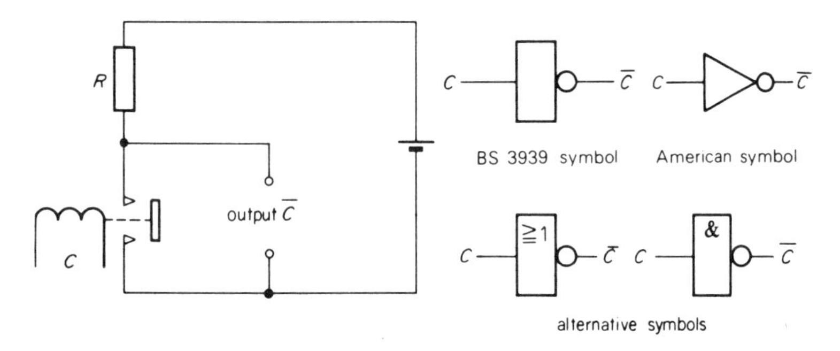


Figure 2.3 A logic NOT gate

One form of relay NOT gate is illustrated in figure 2.3. When $C = 0$, the relay is de-energised and the output voltage is HIGH, that is, $\overline{C} = 1$. When $C = 1$, the relay is energised and the output terminals are short-circuited together so that $\overline{C} = 0$. Resistor $R$ is included in the circuit to limit the current drawn from the supply when the relay contacts are closed.

## 2.4 COMPLETE COIN CIRCUIT OF THE VENDING MACHINE

A block diagram of the logic circuit of the coin section of our vending machine is shown in figure 2.4. Inputs $F$ and $T$ are activated by placing a coin in the appropriate slot in the machine, and signal $C$ is generated by the presence of a 5p coin inside the machine.

When using the block-diagram technique, we only show the

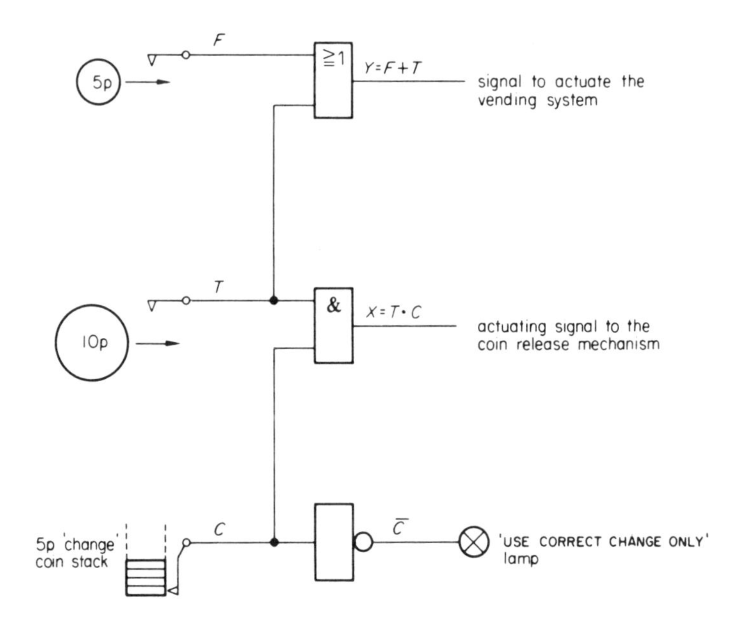


Figure 2.4 A block diagram of the logic circuit for a vending machine

connections through which information signals flow. Connections which are concerned with the power supply, for example, the main supply line, bias supplies and earth line are not shown.

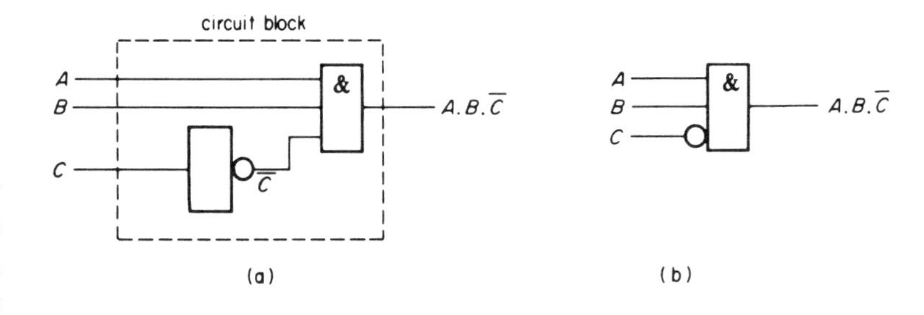


Figure 2.5 A gate with a negated (INVERTING) input

## 2.5 NEGATED INPUTS

Some circuits have a built-in NOT gate associated with individual inputs, shown in the case of input $C$ in figure 2.5a. Where this occurs, the negated input is represented by a circle at the input, shown in figure 2.5b.

## 2.6 THE EFFECT OF POSITIVE AND NEGATIVE LOGIC LEVELS

In the first chapter it was pointed out that a specific electrical voltage may represent either a logic '1' or a logic '0' signal. So far we have talked in terms of a positive voltage representing '1', and zero potential representing '0'. Let us now consider the effect of operating a given logic gate alternately with positive and negative logic levels.

Suppose that we have measured the input and output voltages associated with a two-input gate, and have found them to be as shown in table 2.4, where $H$ is a high voltage and $Z$ is zero voltage.

**Table 2.4**

| *Inputs* | | *Output* |
|---|---|---|
| $A$ | $B$ | $X$ |
| $Z$ | $Z$ | $Z$ |
| $Z$ | $H$ | $Z$ |
| $H$ | $Z$ | $Z$ |
| $H$ | $H$ | $H$ |

### Positive Logic Operation

In the positive logic notation, $H = 1$ and $Z = 0$. If we rewrite the voltage levels in table 2.4 in terms of positive logic levels, we obtain table 2.5.

Comparing table 2.5 with the truth table for a two-input AND gate, table 2.1, we see that when we use the positive logic notation the gate generates the *AND function* of the inputs.

**Table 2.5**

| *Inputs* | | *Output* |
|---|---|---|
| $A$ | $B$ | $X$ |
| 0 | 0 | 0 |
| 0 | 1 | 0 |
| 1 | 0 | 0 |
| 1 | 1 | 1 |

### Negative Logic Operation

Using the negative logic notation, $H = 0$ and $Z = 1$. Rewriting the voltage levels in table 2.4 in terms of negative logic, we obtain the results in table 2.6.

**Table 2.6**

| *Inputs* | | *Output* |
|---|---|---|
| $A$ | $B$ | $X$ |
| 1 | 1 | 1 |
| 1 | 0 | 1 |
| 0 | 1 | 1 |
| 0 | 0 | 0 |

Let us now compare table 2.6 with table 2.2 for a two-input OR gate. Comparing *like* input conditions in both cases, we see that table 2.6 is that of an *OR gate.*

### Summary

Clearly the name we give to a logic gate (defined by its truth table) depends on the logic notation used in the system. As we have seen above, a gate which is described as an AND gate when using the positive logic notation also operates as an OR gate when using negative logic notation. You may also like to show that a positive logic OR gate generates the negative logic AND function.

Positive logic devices and systems will largely be considered in the remainder of the book. Specific reference will be made to circuits using negative logic.

## PROBLEMS

**2.1** Draw up truth tables for AND and OR gates each having four input lines.

**2.2** Waveforms $X$ and $Y$ in figure 2.6 are applied to the input of an OR gate. Draw the waveform for the output $Z$ from the gate.

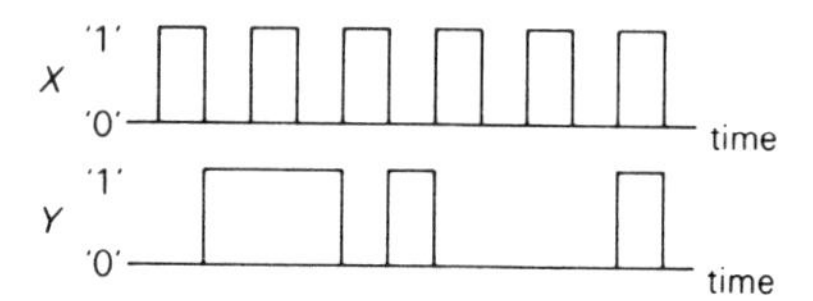


Figure 2.6

**2.3** Draw the output wave form $L$ from a two-input AND gate which has the waveforms $X$ and $Y$ in figure 2.6 applied to its input.

**2.4** Draw the output waveform $M$ from figure 2.7 when the waveforms $X$ and $Y$ in figure 2.6 are applied to the input.

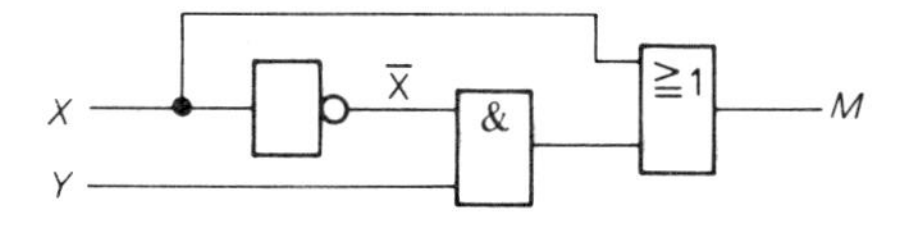


Figure 2.7

**2.5** Determine the logic equation which describes table 2.7.

**2.6** The voltage measured at the input and output of a gate are shown in table 2.8. Give the name of the logic function generated if (a) positive logic is used, (b) negative logic is used.

**Table 2.7**

| *Inputs* | | | *Output* |
|---|---|---|---|
| *A* | *B* | *C* | *K* |
| 1 | 0 | 1 | 0 |
| 0 | 1 | 0 | 0 |
| 1 | 1 | 1 | 1 |
| 1 | 1 | 0 | 0 |
| 0 | 0 | 0 | 0 |
| 0 | 1 | 1 | 0 |
| 1 | 0 | 0 | 0 |
| 0 | 0 | 1 | 0 |

**Table 2.8**

| *Inputs* | | *Output* |
|---|---|---|
| *A* | *B* | *F* |
| −2 V | −2 V | −2 V |
| 6 V | 6 V | 6 V |
| 6 V | −2 V | −2 V |
| −2 V | 6 V | −2 V |

# 3 NAND and NOR Networks

The names NAND and NOR are contractions of the following logic functions

$$\text{NAND} = \text{NOT AND} = \overline{\text{AND}}$$
$$\text{NOR} = \text{NOT OR} = \overline{\text{OR}}$$

The functions are described in detail in the following sections.

## 3.1 THE NAND FUNCTION

The NAND function is generated when an AND gate and a NOT gate are combined in the manner shown in figure 3.1. Its truth table is given in table 3.1.

We shall consider the operation of the circuit in terms of the two stages in figure 3.1. The first stage of the circuit generates the AND

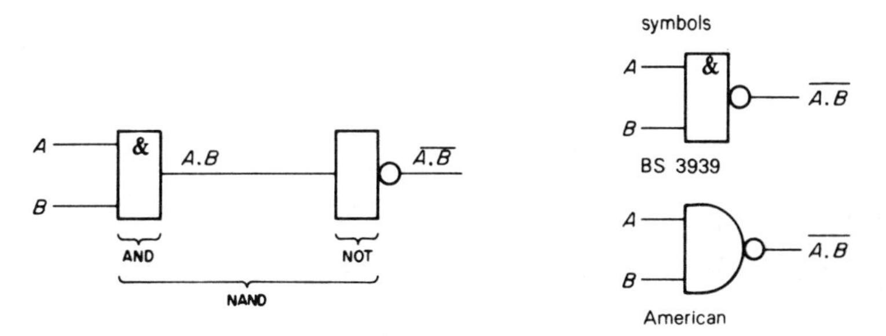


Figure 3.1 A circuit which generates the NAND function

**Table 3.1 Truth table for a two-input NAND gate**

| *Inputs* | | *Intermediate AND function $A.B$* | *Output $A.B$* |
|---|---|---|---|
| 0 | 0 | 0 | 1 |
| 0 | 1 | 0 | 1 |
| 1 | 0 | 0 | 1 |
| 1 | 1 | 1 | 0 |

function which, as we saw earlier, gives an output of '0' whenever any input is '0', and an output of '1' only when both inputs are '1', resulting in the intermediate AND result in table 3.1. The NOT section of the NAND gate complements or inverts the intermediate AND signal to give the final output. The complete truth table can be summarised as follows.

*When any input to a NAND gate is energised by a logic '0' signal, then its output is '1'. Otherwise the output is '0'.*

## 3.2 THE NOR FUNCTION

A block diagram of a network which generates the NOR function is shown in figure 3.2 and has the truth table shown in table 3.2.

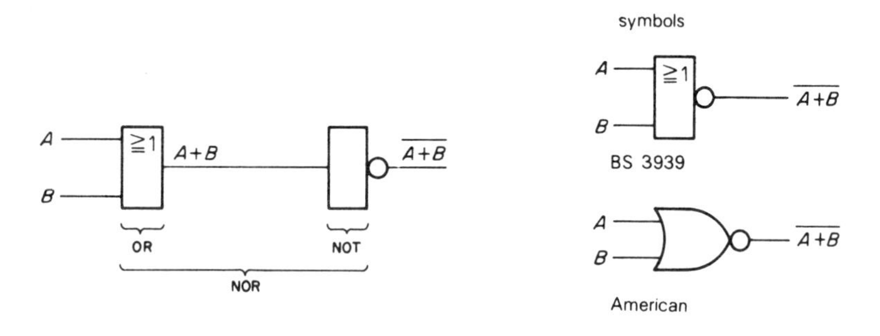


Figure 3.2 A circuit which generates the NOR function

**Table 3.2 Truth table for a two-input NOR gate**

| *Inputs* | | *Intermediate OR function* $A+B$ | *Output* $\overline{A+B}$ |
|---|---|---|---|
| 0 | 0 | 0 | 1 |
| 0 | 1 | 1 | 0 |
| 1 | 0 | 1 | 0 |
| 1 | 1 | 1 | 0 |

The OR section of the circuit generates a logic '1' whenever a '1' signal is applied to either input, and this is inverted by the NOT gate. The overall truth table may be summarised as follows.

*When any input to a NOR gate is energised by a logic '1' signal, then its output is '0'. Otherwise the output is '1'.*

## 3.3 WHY USE NAND AND NOR GATES?

A feature of both NOR and NAND networks *is that they can be used to generate the truth tables of all other types of gates*; that is, a number of NAND gates connected in particular configurations can be used to generate the AND, OR, NOT and NOR functions (see section 3.5), and various types of memory circuits can also be constructed (see chapter 8). This has obvious economic advantages in so far as it is necessary for users to purchase and to stock only one basic type of logic element.

At first sight it may seem that a large number of either NAND or NOR gates are required to replace an equivalent AND–OR–NOT type of network. This, in fact, is not necessarily the case since Boolean algebraic techniques can be used to *minimise* the number of gates required in the system. In many instances it is possible to generate a given logic function using a smaller number of either NAND or NOR gates than is possible with AND, OR and NOT gates. Moreover, the cost of NAND and NOR elements is generally less than that of other types of elements.

## 3.4 THE EFFECT OF POSITIVE AND NEGATIVE LOGIC CONVENTIONS ON NAND AND NOR GATES

We can illustrate the effect of employing either positive or negative logic conventions on the truth tables of these elements by considering table 3.3.

This is a table of input and output voltages of a particular logic gate, where $H$ is a high voltage and $Z$ is zero voltage. In positive logic $H = 1$ and $Z = 0$, and in negative logic $H = 0$ and $Z = 1$. Applying these relationships to table 3.3 we obtain tables 3.4 and

**Table 3.3**

| Inputs | | Output |
|---|---|---|
| $A$ | $B$ | |
| $H$ | $H$ | $Z$ |
| $H$ | $Z$ | $Z$ |
| $Z$ | $H$ | $Z$ |
| $Z$ | $Z$ | $H$ |

**Table 3.4 Positive logic truth table**

| Inputs | | Output |
|---|---|---|
| $A$ | $B$ | |
| 1 | 1 | 0 |
| 1 | 0 | 0 |
| 0 | 1 | 0 |
| 0 | 0 | 1 |

**Table 3.5 Negative logic truth table**

| Inputs | | Output |
|---|---|---|
| $A$ | $B$ | |
| 0 | 0 | 1 |
| 0 | 1 | 1 |
| 1 | 0 | 1 |
| 1 | 1 | 0 |

3.5 for positive logic operation and negative logic operation, respectively.

Comparing table 3.4 with the truth table for a 2-input NOR element, table 3.2, we see that the device may be regarded as a *positive logic NOR gate*. Comparing table 3.5 with the truth table for a 2-input NAND element, table 3.1, we also see that it may be regarded as a *negative logic NAND gate*.

It can also be shown that a *positive logic NAND gate* may be regarded as a *negative logic NOR gate*.

It is for this reason that it is sometimes necessary to state whether positive or negative logic conventions are being used when specifying the name of the logic function generated by the gate. In some cases the gate is described as a NAND/NOR gate and the manufacturer provides the truth table which indicates the voltage levels involved, that is, a truth table similar to table 3.3.

## 3.5 NAND NETWORKS

NAND gates are perhaps the most popular elements in use today and it is possible to generate all the functions so far described by using combinations of these gates. In this section of the book we shall consider how to develop various functions using only NAND gates.

### The NOT Function

The NOT function is generated by the single-input NAND gate in figure 3.3a. You will recall that in the case of a NAND gate, the output is logic '1' if any input is held at logic '0'. Since we have only one input in this case, when $A = 0$ the output is logic '1' ($= \overline{A}$). Also when each input of a NAND gate is logic '1', the output is '0'. Since we have only one input, when $A = 1$ the output is logic '0' ($= \overline{A}$).

Let us consider the operation of the circuit in figure 3.3b. Gate G1 functions as a 3-input NAND gate, whose output is $\overline{A \,.\, B \,.\, C}$. Gate G2 acts as a NOT gate which logically inverts the signal at the output of G1. That is, the output is

$$\overline{\overline{(A \,.\, B \,.\, C)}} = A \,.\, B \,.\, C$$

which is the AND function of the inputs. This is verified in table 3.6.

**Table 3.6 Truth table for figure 3.3b**

| *Inputs* A | B | C | *Intermediate NAND function* $\overline{A.B.C}$ | *Output* = NOT $\overline{A.B.C}$ = $\overline{\overline{A.B.C}}$ = A.B.C |
|---|---|---|---|---|
| 0 | 0 | 0 | 1 | 0 |
| 0 | 0 | 1 | 1 | 0 |
| 0 | 1 | 0 | 1 | 0 |
| 0 | 1 | 1 | 1 | 0 |
| 1 | 0 | 0 | 1 | 0 |
| 1 | 0 | 1 | 1 | 0 |
| 1 | 1 | 0 | 1 | 0 |
| 1 | 1 | 1 | 0 | 1 |

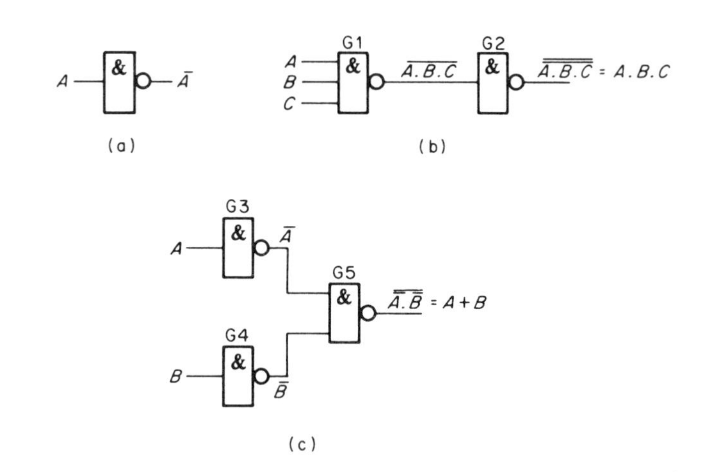


Figure 3.3 Methods of generating (a) the NOT function, (b) the AND function and (c) the OR function using NAND gates only

### The OR Function

In the circuit in figure 3.3c, gates G3 and G4 act as invertors whose outputs are $\overline{A}$ and $\overline{B}$, respectively. Gate G5 operates as a 2-input NAND gate whose inputs are $\overline{A}$ and $\overline{B}$, to give an output of

$$\overline{\overline{A}\,.\,\overline{B}} = A + B$$

which is the OR function of the inputs. This is verified in table 3.7.

**Table 3.7 Truth table for figure 3.3c**

| *Inputs* A | B | *Intermediate NOT gate outputs* $\overline{A}$ | $\overline{B}$ | $\overline{A}\,.\,\overline{B}$ | *Output from G5* = NOT $(\overline{A}\,.\,\overline{B})$ = $\overline{\overline{A}\,.\,\overline{B}}$ = $A+B$ |
|---|---|---|---|---|---|
| 0 | 0 | 1 | 1 | 1 | 0 |
| 0 | 1 | 1 | 0 | 0 | 1 |
| 1 | 0 | 0 | 1 | 0 | 1 |
| 1 | 1 | 0 | 0 | 0 | 1 |

### The NOR Function

Since NOR = NOT OR, this function is generated by driving the input of a NOT circuit of the type in figure 3.3a by an OR circuit of the type in figure 3.3c.

### Basic Minimisation Techniques for NAND Networks

Basic types of NAND networks may require a large number of gates, but it may be possible in many cases to reduce the number of gates required by the application of simple rules.

If two 1-input NAND gates are cascaded in the manner shown in figure 3.4a, the output from G1 is $\overline{A}$ and that from G2 is $\overline{\overline{A}}$ which is equal to $A$. That is *both gates are redundant and can be replaced by a single wire connecting the input to the output.*

In the following, we show that the network in figure 3.4b can be replaced by a 3-input NAND gate. The output from G3 is $\overline{A\,.\,B}$, which is logically inverted by G4, whose output is $A.B$. The

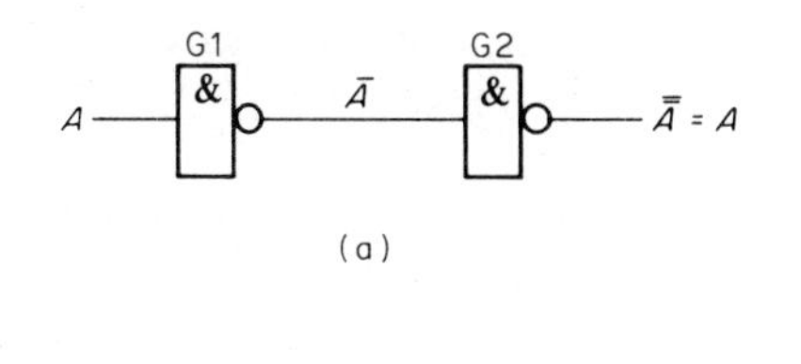


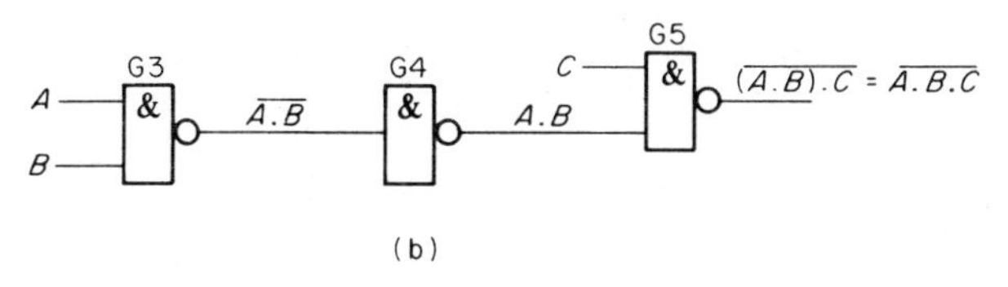


Figure 3.4 Basic minimisation techniques for use in NAND networks

function generated by the 2-input gate G5 is $\overline{(A\,.\,B)\,.\,C} = \overline{A\,.\,B\,.\,C}$. This function would be generated by a single 3-input NAND gate whose inputs are energised independently by signals $A$, $B$ and $C$.

## 3.6 NOR NETWORKS

As with NAND gates, NOR gates can be used to generate the basic logic functions. The block diagrams associated with these networks are given below and it is left as an exercise for you to verify their accuracy using truth tables.

### The NOT Function

We saw when first dealing with the NOR gate that when any input is energised by a logic '1', it causes the output to be logic '0'. Using a 1-input NOR gate, as in figure 3.5a, when the input is logic '0', the output is logic '1'. That is, a 1-input NOR gate acts as a NOT gate.

### The OR Function

In figure 3.5b, the output from G1 is $\overline{A+B}$, and that from G2 is $\overline{\overline{A+B}} = A+B$, that is, it is the OR function of the inputs.

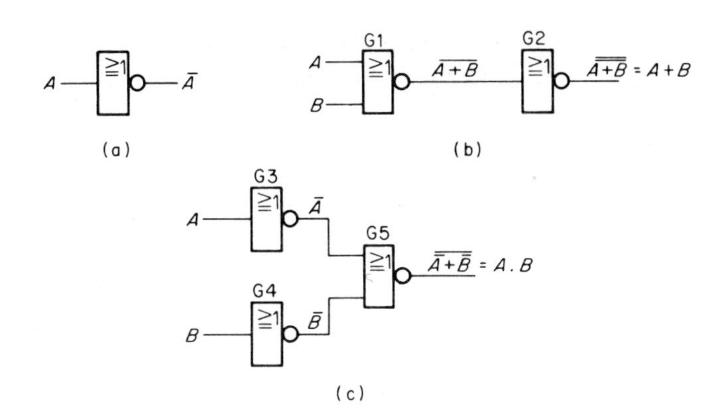


Figure 3.5 Methods of generating (a) the NOT function, (b) the OR function and (c) the AND function using NOR gates only

### The AND Function

For two input signals, this function is developed by figure 3.5c. Gates G3 and G4 act as invertors to give signals $\overline{A}$ and $\overline{B}$ respectively. The output from G5 is $\overline{\overline{A}+\overline{B}}$ which is the AND function of the input signals.

### The NAND Function

Since NAND = NOT AND, we can develop this function by using an AND gate of the type in figure 3.5c to drive the input of a NOT gate similar to that in figure 3.5a, the output of which is the NAND function of the inputs to the combined circuit.

### Basic Minimisation Techniques for NOR Networks

The minimisation techniques developed here are similar to those used earlier with NAND gates.

In figure 3.6a, the double inversion of the single input signal results in both gates being redundant and they can be replaced by a wire linking the input to the output.

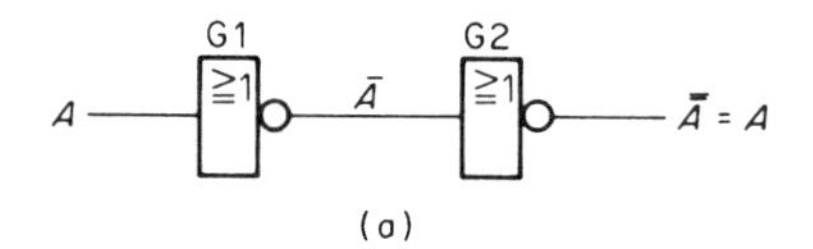


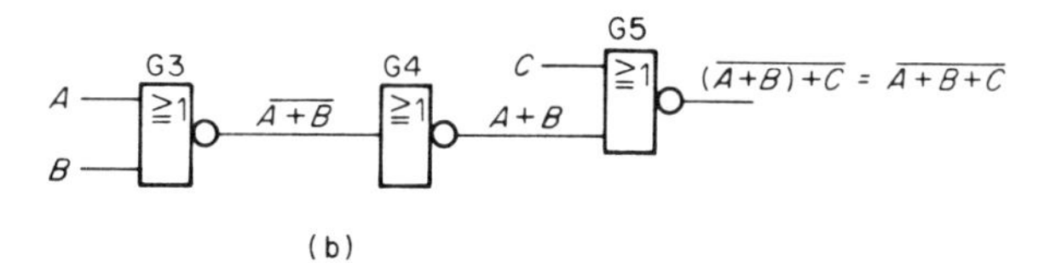


Figure 3.6 Basic minimisation techniques for use in NOR networks

In figure 3.6b, G3 and G4 form an OR gate (see also figure 3.5b), so that the output from G5 is $\overline{(A+B)+C} = \overline{A+B+C}$. That is, the complete circuit in figure 3.6b can be replaced by a 3-input NOR gate.

## PROBLEMS

**3.1** The voltages measured at the input and output of a gate are listed in table 3.8. Determine the logic function which describes the table if (a) positive logic is used, (b) negative logic is used, (c) positive logic is used at the input and negative logic at the output, (d) negative logic is used at the input and positive logic at the output.

**Table 3.8**

| Inputs | | Output |
|---|---|---|
| *A* | *B* | *G* |
| 2 V | 2 V | 2 V |
| 2 V | 6 V | 6 V |
| 6 V | 2 V | 6 V |
| 6 V | 6 V | 6 V |

**3.2** Draw up a truth table for (a) a NOR gate and (b) a NAND gate each having four input lines.

**3.3** Prove that the circuit in figure 3.7 can be replaced by a single NAND gate having four input lines.

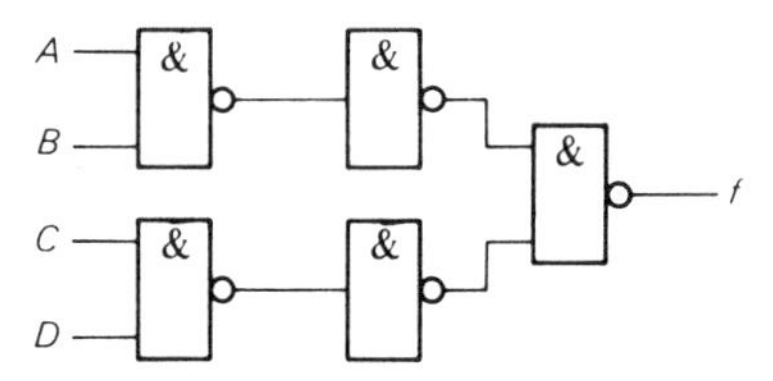


Figure 3.7

**3.4** Name the two-input gate whose input and output waveforms correspond to those in figure 3.8.

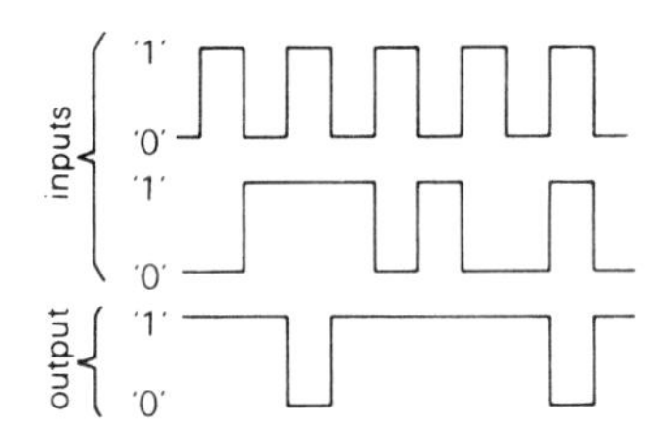


Figure 3.8

# 4 Semiconductor Devices

An ideal switch is one which when it is 'on', has no effective resistance between its terminals and no voltage appears across it when current flows through it. When it is 'off', it has an infinite resistance between its terminals and the leakage current through it is zero. Moreover, our ideal switch has a perfect switching action so that the time taken for the circuit current either to be cut off or to be switched on is zero.

No practical device can hope to attain these ideals, but semiconductor diodes and transistors most nearly approach them.

## 4.1 SEMICONDUCTOR MATERIALS

Transistors and diodes are manufactured from a range of materials known as *semiconductors*, and have certain valuable properties which allow them to be used to control the flow of current in electronic circuits. A semiconductor is a material whose conductance at room temperature is mid-way between that of a good conductor and that of a good insulator. In quantitative terms, the conductance of semiconductors lies in the range $10^{-4}$ Ωm to $10^{3}$ Ωm.

The two principal semiconductor materials used for the construction of transistors and diodes are *silicon* (Si) and *germanium* (Ge), the former being universally used in electronic logic devices. In the early years of semiconductor electronics, germanium was widely used, since the production technology associated with it was well understood. Silicon has a number of advantages in switching applications over germanium, including a wider operating temperature range and a lower leakage current. Silicon production techniques have overtaken those of germanium, and present generations of silicon devices have superior performance parameters than those of germanium devices. Other types of semiconductor materials, for example, gallium arsenide (GaAs), are used in specialised applications.

Semiconductors can exist in three basic forms, which are *i-type*, *p-type*, and *n-type*. *Intrinsic* semiconductors (*i*-type) are commercially pure semiconductors. The conductivity of this type of material increases with temperature; that is, it has a *negative* resistance–temperature coefficient and the flow of current through it is largely temperature dependent. Ideally, *i*-type semiconductors should be good insulators at room temperature, silicon approaching this more closely than germanium.

Electrons in the *outermost* orbit of atoms are known as *valence electrons* and are responsible for the chemical and electrical properties of the atom. Silicon and germanium atoms have four valence electrons and are known as *tetravalent* atoms. Electrical conduction occurs as a result of the movement in an electric field of valence electrons which have become detached from parent atoms. Since the negative charge of the valence electrons ($-4$ units of charge for a tetravalent atom) just balances the positive charge ($+4$ units) on the remainder of the atom, the *net* electrical charge on an isolated atom is zero. When an electron leaves the atom, it takes with it a negative charge of $-1$ unit and leaves behind an atom with a *net* positive charge of $+1$ unit on it. This positive charge is described as a *hole*, and is equivalent to an electron deficiency. Thus current flow is due to both the flow of electrons towards the positive pole of the supply and of holes towards the negative pole. The type of charge carrier primarily involved in the conduction process depends on the type of semiconductor in use.

In *i*-type semiconductors, the valence electrons are all required for chemical bonding purposes, and for this reason *i*-type materials are poor conductors.

If we deliberately introduce a known amount (about 1 part in $10^8$) of a specific type of impurity, the electrical properties of the semiconductor are altered in a particular way. If, for example, we introduce an impurity atom which has five valence electrons (*pentavalent atoms*) into silicon or germanium, only four of the five valence electrons are required for chemical bonding purposes. The remaining electron can easily be detached for conduction purposes, so that this electron becomes a *mobile charge carrier* within the atomic structure. Since the mobile charge carrier is an electron, it has a negative charge on it and the material is known as an *n-type* semiconductor. In *n*-type materials, since electrons are mobile, *current flow is due primarily to the movement of negative charge carriers from the negative pole of the supply to the positive pole.*

Pentavalent impurity substances include arsenic (As), phosphorus (P) and antimony (Sb).

On the other hand, if we introduce an impurity atom which has three electrons in its outer orbit (*trivalent atoms*), such as aluminium (Al), boron (B), gallium (Ga) or indium (In), the resulting material has an electron deficiency in its structure. That is, *conduction takes place as a result of the movement in an electric field of mobile positive charge carriers or holes.* This type of material is known as a *p-type semiconductor.*

## 4.2 SEMICONDUCTOR JUNCTION DIODES

The simplest type of electronic switch is the semiconductor diode, shown in figure 4.1. This device is constructed in a single crystal of

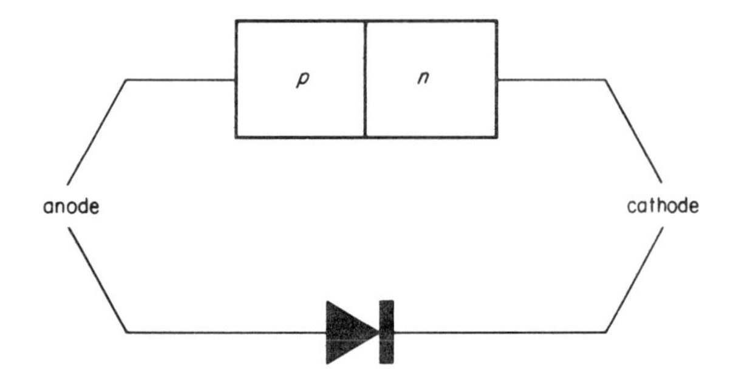


Figure 4.1 A *p–n* junction diode

semiconductor which contains a *p*-region and an *n*-region. The *p*-region is the *anode* of the diode and the *n*-region is its *cathode*, and when the anode is positive (*p-region positive*) with respect to the *n*-region, the diode is said to be *forward biased* and current flows through it. Under these conditions the *forward voltage drop* across the diode when carrying its rated current is about 0.7–0.8 V in the case of a silicon device and about 0.3–0.4 V in the case of a germanium diode. The diode is said to be in its *forward conduction mode*, corresponding to operation in the first quadrant of the characteristic in figure 4.2. In this mode, the diode operates as a switch which is turned *on.*

If we connect the anode to the negative pole of the supply and the cathode to the positive pole, the diode cuts off the flow of

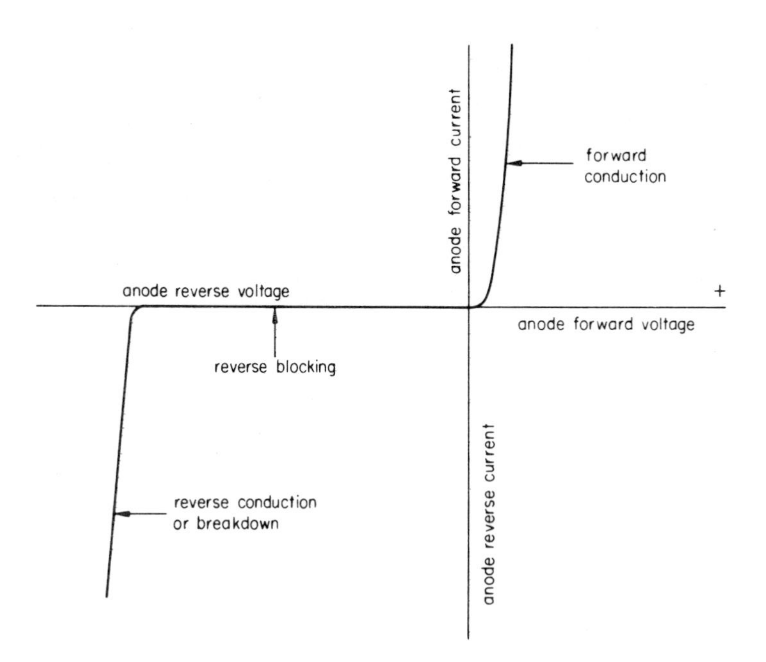


Figure 4.2 Characteristic of a *p–n* junction diode

current and is then said to be in its *reverse blocking mode*. Providing that the *reverse voltage* or *inverse voltage* applied to the diode does not exceed its *breakdown voltage*, only *leakage current* flows between the two regions. The leakage current in silicon devices may be as low as a few nanoamperes (1 nA = $10^{-9}$ A). In this mode the diode acts as though it were a switch which is turned *off*.

If the reverse breakdown voltage is exceeded, the reverse current increases at a rapid rate. With conventional diodes this operating region is avoided, since it can rapidly lead to excessive heating with consequent damage to the diode.

Certain types of diode, known as *Zener diodes*, are designed to operate in the reverse breakdown mode. The symbol for the Zener diode is shown in figure 4.3. Zener diodes are used in a wide range of circuits including bias circuits, voltage offset circuits, voltage reference circuits, and stabilised power supplies.

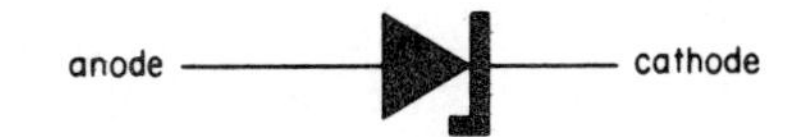


Figure 4.3 Circuit symbol for a Zener diode

## 4.3 CHARGE CARRIER STORAGE IN DIODES

When the bias applied to a junction diode changes from forward bias to reverse bias, the charge carriers within the device must recombine with parent atoms and disappear before the current through the diode can fall to zero. This action results in a pulse of reverse current flowing through the diode, in the manner shown in figure 4.4. Here the circuit is limited in magnitude to $V/R$ by the circuit resistance. The effect which causes the reverse current is known as *charge storage*; the net effect of charge storage is to introduce a delay between the time that the supply voltage is reversed and the time that the anode current falls to zero (or, more accurately, to its leakage value).

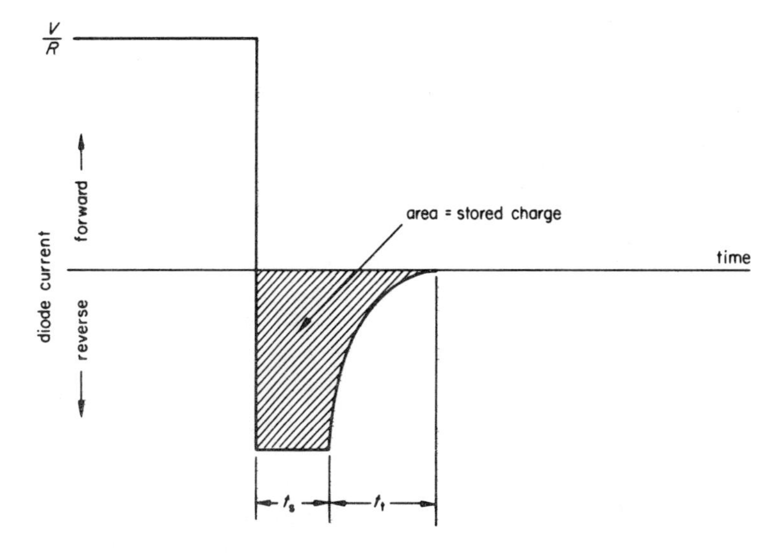


Figure 4.4 Charge storage effect

The *storage time* $t_s$ of the diode is the time taken for the reverse voltage to sweep the charge carriers away from the junction; the *transition time* $t_t$ is the time taken for the reverse current to fall to zero. The total delay is sometimes quoted as the *recovery time* of the diode and, in switching diodes, can be as small as a few nanoseconds (1 ns = $10^{-9}$ s). Diodes used in switching circuits should have as small as possible a recovery time.

Charge storage has a limiting effect on the ultimate switching speed of all semiconductor devices employing *p–n* junctions whose bias is reversed from time to time. The bipolar junction transistor (see section 4.5) is no exception to this.

## 4.4 THE SCHOTTKY BARRIER DIODE

The Schottky barrier diode is a device with a metal-to-semiconductor (usually *n*-type) rectifying junction. The operating principle of this device differs from that of the *p–n* junction diode, and it does not exhibit the storage time delay associated with it. The latter factor makes the Schottky barrier diode an attractive device for some applications in switching circuits (see also section 5.14).

## 4.5 THE BIPOLAR JUNCTION TRANSISTOR (BJT)

The bipolar junction transistor is a device constructed in a single crystal of semiconductor material, and has three regions known respectively as the *emitter*, the *base* and the *collector*. Figure 4.5 shows a pictorial representation of the two main types of bipolar transistor, known as *n–p–n* and *p–n–p* transistors respectively. During the process of conducting current, both types of BJT employ both holes and electrons in their operation, and it is for this reason that they are described as *bipolar* devices.

In both types of transistor the emitter is the region which emits charge carriers into the transistor, the base is the region used to control the flow of current through the device and the collector is the region in which the charge carriers are collected.

The manner in which the base current controls the collector

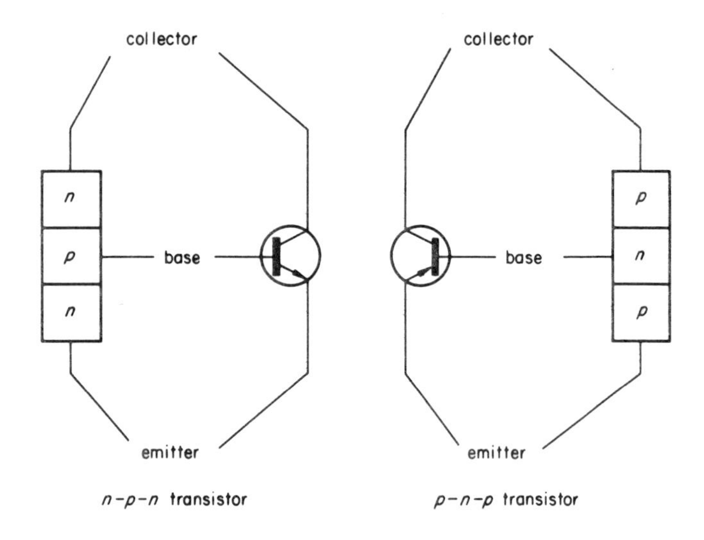


Figure 4.5 Circuit symbols for bipolar junction transistors

current is best illustrated by means of the *output characteristic* or *collector characteristic* of the transistor, an example of which is shown in figure 4.6a. The characteristics shown are the *common-emitter characteristics* of an *n–p–n* transistor and are obtained by testing the transistor with the emitter region used as the connection that is common to both the base signal (that is, the control signal) and the collector supply voltage. A typical test circuit used to obtain these characteristics is shown in figure 4.7. In switching circuits, BJTs are almost invariably used in this operating mode (known as the *common-emitter mode*) since it offers both a high current gain and a high power gain.

We see from figure 4.6a that when the base current $i_B$ is zero, the collector current is also zero. In fact this is not quite true, since a small value of leakage current [usually in the range between a few nanoamperes (nA) and a few microamperes (μA)] does flow. When $i_B = 0$, the transistor is said to be *cut off*, and it operates as a switch which is *off*. As the base current is increased, we find that the collector current also increases. With a base current of 75μA we see that the collector current at point A is 7.4 mA and a base current of 100 μA gives a collector current at point B of 10 mA.

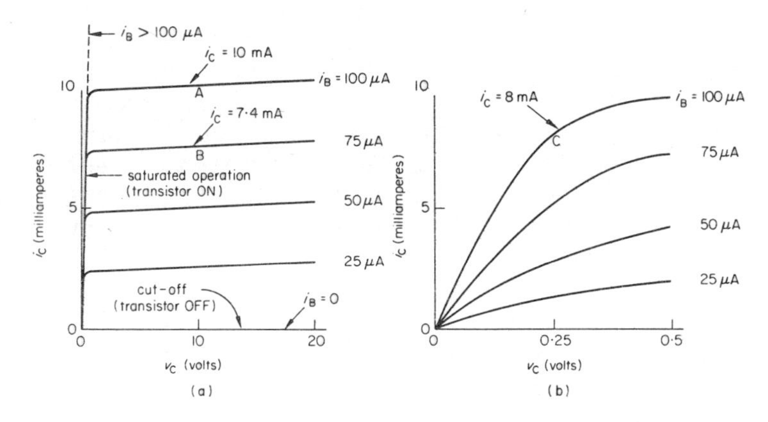


Figure 4.6 (a) Common-emitter collector characteristics for an *n–p–n* transistor and (b) the collector characteristic in the region of the origin

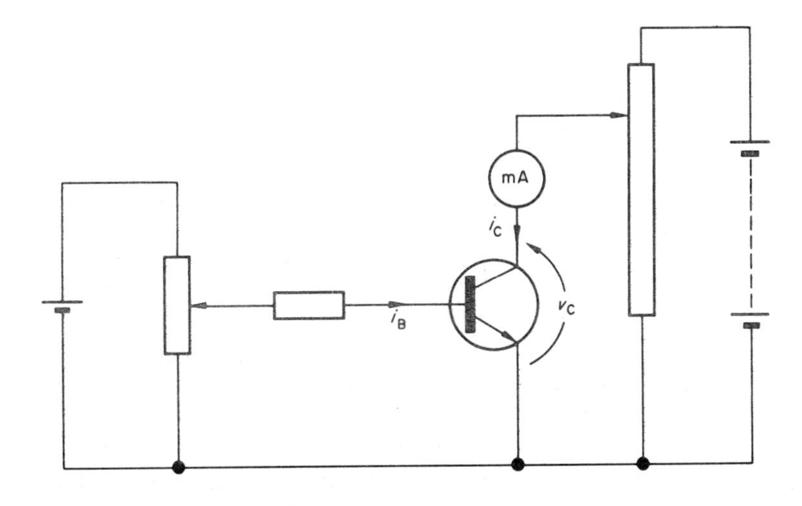


Figure 4.7 A circuit used to determine the common-emitter output characteristic

An important parameter of the BJT is its *forward current gain*, which is defined under a variety of conditions discussed below. The absolute current gain or *d.c. forward current gain*, designated the symbol $h_{FE}$, is the ratio of the total collector current to the total base current. At the point A on figure 4.6a

$$h_{FE} = 10\ \text{mA}/100\ \mu\text{A} = 100$$

This parameter is of particular significance when bias conditions are being evaluated. The *small-signal forward current gain* $h_{fe}$ is the ratio of the *change* occurring in the collector current when the base current is changed in value; this ratio is determined at a constant value of collector voltage (10 V in figure 4.6a). The value of $h_{fe}$ determined between the points A and B on figure 4.6a is

$$h_{fe} = (10 - 7.4)\ \text{mA}/(100 - 75)\ \mu\text{A} = 104$$

Parameter $h_{fe}$ is of value in the design of small-signal linear amplifiers.

Of particular importance in switching circuit design is the *saturated forward current gain*, designated the symbol $h_{FE(sat)}$. A BJT is said to be saturated when its operating point lies on the steep part of the curves on the extreme left-hand of figure 4.6a. When in the saturated state, the transistor carries a large current and supports only a small voltage across it. This section of the characteristics is expanded in figure 4.6b. The value of $h_{FE(sat)}$ evaluated at point C at a collector voltage of 0.25 V is

$$h_{FE(sat)} = 8\ \text{mA}/100\ \mu\text{A} = 80$$

Thus, to drive the transistor into saturation, that is, to cause it to operate as a switch in the *on* condition, the base current must be *at least* 1/80th of the maximum collector current. In circuits which use saturated transistors as switches, the base current used to saturate the transistors is in excess of the minimum saturation value.

When the transistor saturates, the *collector–emitter saturation voltage*, $V_{CE(sat)}$, is very small, its value being 0.25 V in figure 4.6b. Another parameter of importance is the voltage applied to the base–emitter junction in order to saturate the transistor. This parameter is designated the symbol $V_{BE(sat)}$, and has a value of about 0.7–0.8 V in silicon transistors and about 0.4 V in germanium transistors.

## 4.6 INSULATED-GATE FIELD-EFFECT TRANSISTORS

Insulated-gate field-effect transistors (IGFETs) depend for their operation on the influence of an electric field on the conductivity of a very thin region of semiconductor material known as a conducting *channel*. A cross-section through an *insulated gate p-channel FET* is shown in figure 4.8a. The *drain electrode* is maintained at a negative potential with respect to the *source electrode* so that, with

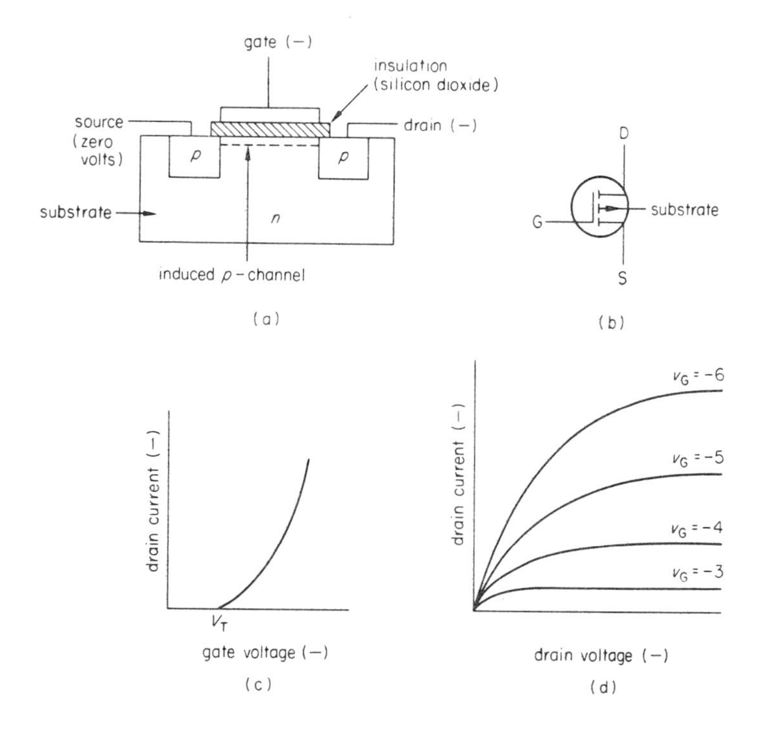


Figure 4.8 (a) Sectional view of a *p*-channel insulated-gate FET, (b) its circuit symbol, (c) typical gate characteristics and (d) typical common-source characteristics

zero gate voltage, the *p–n* junction between the drain and the substrate is reverse biased. The initial drain current is therefore zero, and this lack of conductivity is represented in the circuit symbol, figure 4.8b, by breaks in the link between the source (S) and the drain (D). The construction of the device also leads to it being described as a *MOSFET* (Metal Oxide Semiconductor FET), since the gate-to-channel construction has a metal-oxide semiconductor geometry.

The application of a negative potential to the gate attracts any free *p*-type charge carriers (that is, holes) in the substrate to the underside of the oxide immediately below the gate. At a voltage known as the *threshold voltage*, $V_T$, a sufficient number of positive charge carriers have collected at the oxide-to-semiconductor interface to form a conducting channel which links the source and drain. The value of $V_T$ lies, typically, between about −2 V and −5 V. This is shown in the mutual characteristic of the MOSFET in figure 4.8c. Increasing the negative gate voltage causes the conductivity of the channel to increase, so that the drain current increases in the manner shown in figure 4.8d.

Several types of FET are manufactured, one type for use in logic circuits being the *p*-channel MOSFET described above. The principal features which make MOS devices more attractive in some instances than bipolar transistors in logic applications are listed below.

(1) MOS devices are physically smaller than bipolar transistors, giving a reduction in the cost per gate.

(2) Their construction is compatible with monolithic integrated circuit production techniques.

(3) Their input resistance is very high, typically $10^{12}$ Ω.

(4) MOS devices can be used to replace resistors, with a saving both in size and cost. In this mode, MOS devices are described as *pinch-effect resistors*.

(5) The output from the driving gate can be directly connected to the inputs of the driven gates without need for bias networks.

A disadvantage of MOS devices when compared with bipolar

devices is their lower switching speed. This is due largely to the input capacitance associated with the oxide layer which separates the gate from the conducting channel.

MOSFETs do not exhibit the charge storage effects associated with BJTs, but they do display time lags due to interelectrode capacitance.

## PROBLEMS

**4.1** Explain why the intrinsic conductivity of a semiconductor material increases with increase in temperature.

**4.2** If the resistivity of silicon is 2000 Ωm at 25 °C and if the intrinsic conductivity increases by 7 per cent for each °C rise in temperature, determine its resistivity when the temperature (a) is reduced by 10 per cent, (b) is 27 °C.

**4.3** Explain what is meant by the following terms in connection with semiconductors: majority charge carrier, *p*-type, hole–electron pair, pentavalent atom.

**4.4** Explain why a piece of *n*-type material is electrically neutral. Why does a depletion region appear between the *n*-region and the *p*-region in a diode before the supply is connected?

**4.5** What is meant by the following terms in connection with a diode: forward bias, reverse blocking, reverse breakdown, charge carrier storage.

**4.6** Why does a reverse biased diode possess capacitance between the anode and the cathode? Does this capacitance (a) increase or (b) decrease when the reverse bias voltage is increased? Give reasons for your answer.

**4.7** Why are the *h*-parameters widely used in connection with bipolar junction transistors?

**4.8** Explain the changes which take place on the output characteristics of a bipolar junction transistor when its temperature is raised.

# 5 Semiconductor Logic Circuits and Integrated Circuits

In this chapter we shall deal with all the popular digital logic families and shall outline the important constructional techniques associated with their production. The principal logic families are

| | |
|---|---|
| DRL | diode–resistor logic |
| RTL | resistor–transistor logic |
| DTL | diode–transistor logic |
| TTL | transistor–transistor logic |
| $I^2L$ or MTL | integrated injection logic or merged–transistor logic |
| ECL | emitter-coupled logic |
| MOS and CMOS | MOS logic families |

## 5.1 DIODE–RESISTOR LOGIC (DRL)

DRL is a basic logic family and, as a group of circuits, plays little part in logic systems as we know them today. However, the principles involved in their operation are of great importance to the *integrated circuits* (*IC*) which follow.

### OR Gate

A basic 2-input OR gate circuit is shown in figure 5.1, consisting of two diodes and a resistor. With the input switches in the positions

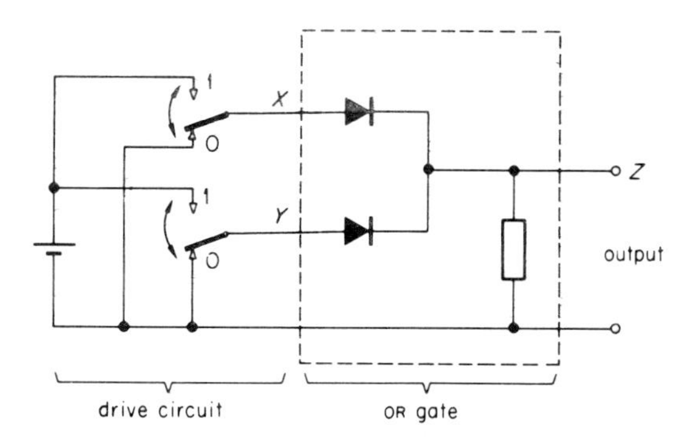


Figure 5.1 A DRL OR gate

shown, that is, $X = Y = 0$, the net e.m.f. acting around the circuit is zero, so that $Z = 0$.

When either $X = 1$ or $Y = 1$, or when both inputs are at the logic '1' level, one or both of the diodes are forward biased so that $Z = 1$. This operation satisfies the OR function truth table.

### AND Gate

In figure 5.2, when either input is switched to the logic '0' position, the diode in that line is forward biased, thereby connecting the output line to zero potential. That is, when either input is logic '0', the output is also logic '0'. Only when both inputs are at the logic '1' level are the diodes cut off. When the circuit is in this state, the current in $R$ falls to a low value and the output voltage rises to the logic '1' level.

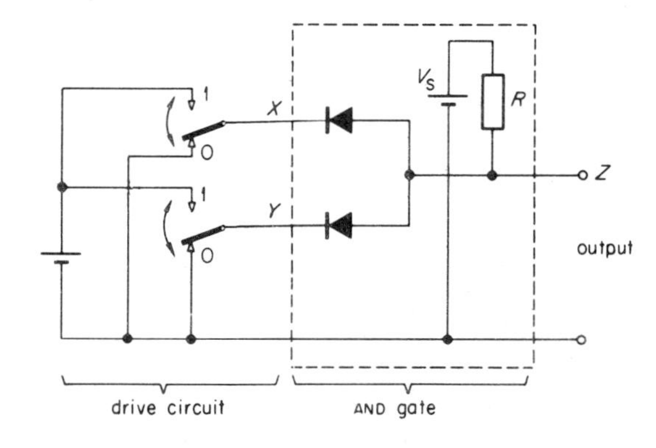


Figure 5.2 A DRL AND gate

## 5.2 LIMITATIONS OF DRL GATES

A primary limitation of DRL gates arises from the fact that diodes are not ideal switches. That is, not only is a potential dropped across each diode when it conducts, but also a current leaks through it when it is reverse biased or is cut off.

Suppose that the voltage supply to the logic system is +5 V, and that the p.d. across each diode is 0.6 V when forward biased. In the case of the OR gate in figure 5.1, when input $X$ is logic '1', the output voltage is given by (5 – forward p.d. across one diode) = 4.4 V (this also applies when input $Y$ is logic '1'). If output $Z$ acts as an input to another OR gate, there is a further 0.6 V drop in the diode in the second gate, resulting in an output voltage from the second gate of 3.8 V. Thus we see that when a number of OR gates are cascaded, the voltage level associated with a logic '1' at the system input is progressively reduced. If sufficient gates are cascaded, the output voltage falls to a value where the system cannot distinguish between logic '1' and logic '0'.

For the above reason, the logic '1' level may be specified, for example, as being in the range +3 to +5 V. Also, for reasons given below, logic '0' may be specified as being a voltage in the range, say, 0 to +0.7 V. In a 'healthy' system, the voltage levels should fall into one of the two above bands. Any voltage in the range 0.7 V to 3 V is a 'false' level and may result in the logic system identifying the signal incorrectly.

In the case of the AND gate in figure 5.2, when input $X$ is logic '0' (shown in the figure) the output voltage at $Z$ is equal to the 'forward' p.d. across the diode (this also applies when $Y = 0$). That is, when $X = 0$ V then $Z = 0.6$ V. The forward p.d. across individual diodes may differ from this value by a small amount, so that the system recognizes a logic '0' as any voltage in the range from zero to about +0.7 V.

Another disadvantage of DRL gates when compared with other logic families is the long propagation delay suffered by signals transmitted through them. This is due not only to the storage time associated with diodes (see section 4.3), but also to the time delay arising from stray capacitances in the circuit. These effects are increased in large systems and result in a relatively slow operating speed when compared with modern systems.

## 5.3 CURRENT-SOURCING AND CURRENT-SINKING

A *current-sourcing* logic gate is one in which the output of a *driving gate* acts as a current source for the current which flows *into* the input of *driven gates* (see figure 5.3). Note that in this diagram, both driving and driven gates are OR-type gates similar to that in figure

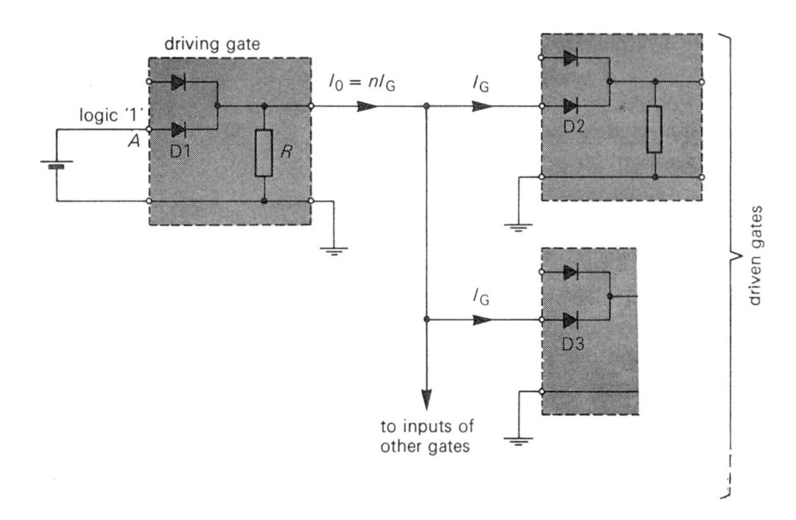


Figure 5.3 Current-sourcing gates

5.1. When the signal applied to input $A$ is logic '1', diode D1 is forward biased and operates as a current source to turn *on* diodes D2, D3, etc., in the driven gates. Thus in figure 5.3 $I_0 = nI_G$

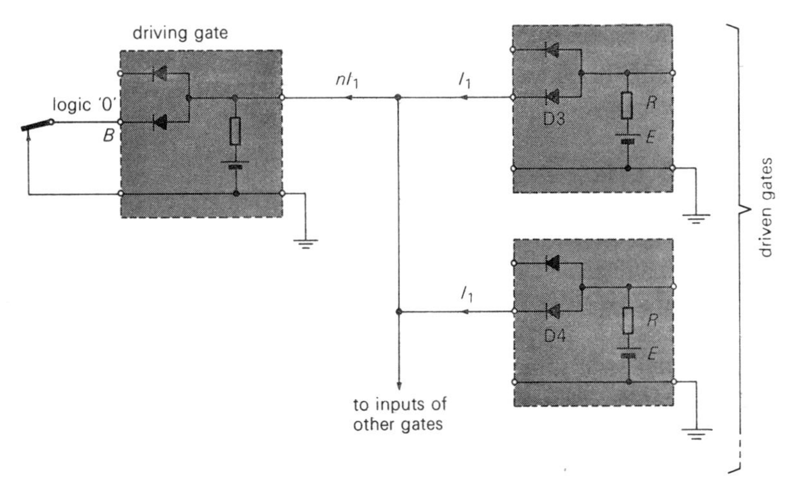


Figure 5.4 Current-sinking gates

where $I_0$ is the output current from the driving gate, $n$ is the number of driven gates, and $I_G$ is the current drawn by one input in a driven gate.

A *current-sinking* logic gate is one in which the output of the driving gate must absorb or *sink* current which flows *out* of the 'input' terminals of the driven gates (see figure 5.4). In the circuit shown, AND-type gates similar to those in figure 5.2 are involved. When the signal applied to input $B$ is logic '0', the current flowing out of the diode in the input line of each driven gate is $I_1$. If there are $n$ driven gates, then the driving gate must 'sink' $nI_1$. In general, gates having AND-type inputs (including NAND gates) are current-sinking gates.

There are exceptions to the above rules, in which a given type of gate may act either as a current-sourcing gate or as a current-sinking gate, depending not only on the type of load but also on its method of connection.

## 5.4 FAN-OUT AND FAN-IN OF LOGIC GATES

As described in section 5.2, the voltage associated with a logic '1' signal at the output of a gate is generally less than that of the system supply voltage. Also, if a number of DRL gates are cascaded, the p.d. in the gates causes an output voltage corresponding to logic '1' to gradually diminish as it progresses through the system.

A similar effect arises when a large number of driven gates are connected to the output of a single driving gate as follows. Consider the current-sourcing circuit in figure 5.3. Suppose that the system supply voltage is 10 V, that the acceptable voltage band for a logic '1' signal is 6–10 V and that the value of the resistor $R$ in each gate is 1 kΩ. If the output resistance of the signal source is 100 Ω, then the voltage $V_0$ appearing between the output terminals of the gate can be calculated from the expression

$$V_0 = \frac{R_L}{100 + R_L} \times 10 \text{ V}$$

where $R_L$ is the effective resistance connected between the output terminals of the logic signal source. If $n$ is the number of driven gates, then when $n = 0$ (that is when no other gates are connected), the effective value of $R_L$ is the 1 kΩ resistor inside the gate itself (see figure 5.5). In this case the output voltage from the gate with 10 V applied is calculated to be 9.1 V. When one gate ($n = 1$) is connected to the output terminals, the effective value of $R_L$ is 500 Ω (equivalent to two 1 kΩ resistors in parallel with one another). This gives an output voltage of 8.33 V. The output voltage for various values of $n$ is listed in table 5.1.

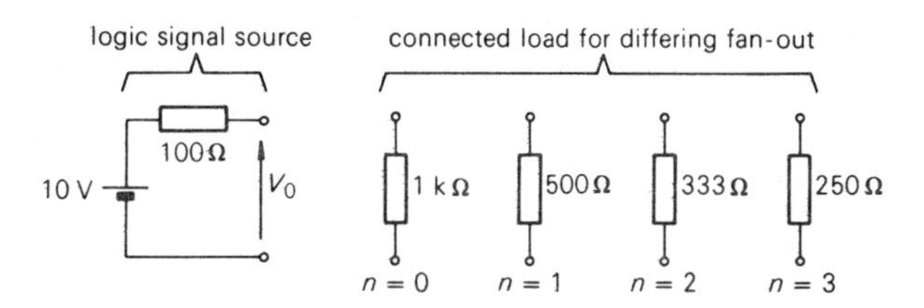


Figure 5.5 The effect of fan-out on a current-sourcing gate

**Table 5.1**

| *Number of connected gates, n* | *Output voltage $V_0$* |
|---|---|
| 0 | 9.1 |
| 1 | 8.33 |
| 2 | 7.7 |
| 3 | 7.14 |
| 4 | 6.7 |
| 5 | 6.25 |
| 6 | 5.88 |

Since the specified range of output voltage corresponding to logic '1' is 6–10 V, the maximum number of gates which may be connected is 5. This is known as the *fan-out* of the gate, and is the maximum number of 'unit' loads which may be connected and still allow the system to operate within its specified logic voltage range.

In practice the situation is worse than suggested above, since we have neglected the effect of the p.d. across the diode in the gate. This may result in the fan-out being reduced to 4 or even to 3.

The above description refers to a current-sourcing gate. Other restrictions apply to current-sinking gates, one being illustrated below. If the maximum current that the output of a gate can 'sink' is 20 mA and if the current flowing from the input line of each driven gate is 2 mA, then the fan-out is limited to 10.

The *fan-in* of a logic circuit is the maximum number of input lines which may be connected to the gate; the gates in figures 5.1 and 5.2 are 2-input gates.

In practice there is a trade-off between fan-out, fan-in and switching speed, the relationship usually being fairly complex. Thus the switching speed is improved slightly if the fan-out and/or fan-in is reduced and vice versa.

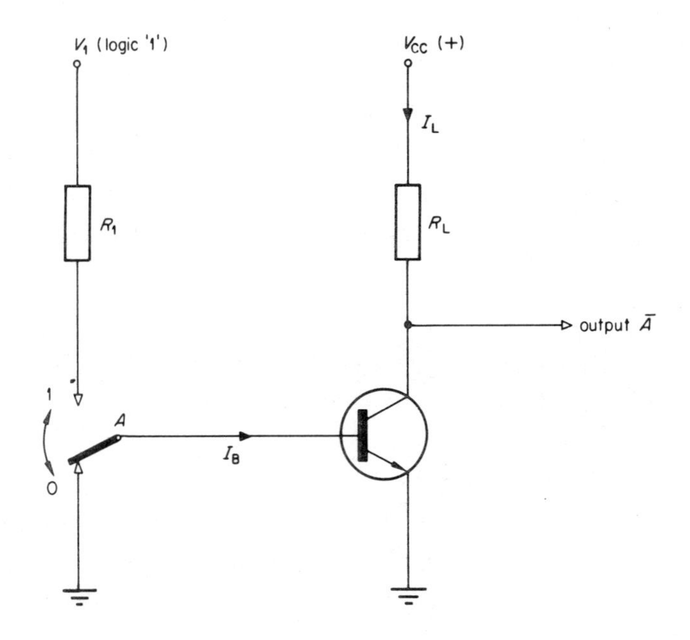


Figure 5.6 An RTL NOT gate

## 5.5 A RESISTOR–TRANSISTOR LOGIC (RTL) NOT GATE

A basic form of RTL NOT gate is shown in figure 5.6. Input $A$ applied to the gate is obtained from a switch which connects the base of the transistor either to earth (logic '0') or to positive voltage $V_1$ (logic '1'). When $A = 0$, as shown in the diagram, the base current is zero and the transistor is cut off. In this case, the collector current is zero and the output signal ($\overline{A}$) is logic '1'. If the output is unloaded, $I_L = 0$ and the output voltage is equal to $V_{CC}$. When the output is loaded, the current flowing in $R_L$ causes the output voltage to fall below $V_{CC}$, and it is not unusual to design circuits to operate with a logic '1' level in the range 0.5 $V_{CC}$ to $V_{CC}$.

When input $A$ is switched to $V_1$, that is, to logic '1', the current flowing in $R_1$ drives the transistor into saturation, so that output $\overline{A}$ is logic '0'.

We shall now consider the design of an RTL NOT gate, the design procedure producing a circuit which will operate satisfactorily in an *unloaded* state. The design is then extended to allow for the effects of loading and other factors.

When the transistor is saturated, the current $I_{L(sat)}$ flowing in the load is

$$I_{L(sat)} = \frac{V_{CC} - V_{CE(sat)}}{R_L} \approx \frac{V_{CC}}{R_L}$$

The simplification in the above relationship is brought about by the assumption that $V_{CE(sat)}$ is much smaller than $V_{CC}$. The base current $I_{B(sat)}$ needed to saturate the transistor is

$$I_{B(sat)} = \frac{I_{L(sat)}}{h_{FE(sat)}} \approx \frac{V_{CC}/R_L}{h_{FE(sat)}}$$

where $I_{L(sat)}$ is the saturation current flowing through the transistor. Now

$$I_{B(sat)} = \frac{V_1 - V_{BE(sat)}}{R_1} \approx \frac{V_1}{R_1}$$

Solving for $R_1$ between the two equations for $I_{B(sat)}$ yields

$$R_1 = \frac{(V_1 - V_{BE(sat)})\, h_{FE(sat)} R_L}{V_{CC} - V_{CE(sat)}}$$

or

$$R_1 \approx \frac{V_1}{V_{CC}} h_{FE(sat)} R_L$$

As shown earlier, the voltage representing logic '1' may have any value between $V_{CC}$ and $V_{CC}/2$, or even lower. Thus the *minimum* value of $R_1$ is needed when $V_1 = V_{CC}/2$, that is when $R_1 = R_L h_{FE(sat)}/2$. When $V_1 = V_{CC}$, it is possible to use a value of $R_1 = R_L h_{FE(sat)}$. The former value is chosen since it allows the transistor to saturate when the lowest logic '1' signal is applied. In practice, it is also necessary to allow for variations not only in the supply voltage, but also in the tolerances of $R_1$ and $R_L$ as well as changes in $h_{FE(sat)}$ between transistors. To allow for all these factors, the value chosen for $R_1$ may only be about 30 per cent of the maximum value, that is, 30 per cent of $R_L h_{FE(sat)}$. To illustrate the design procedure, we shall design a NOT gate which uses a supply voltage of 10 V and in which the maximum collector current is to be 5 mA and the value of $h_{FE(sat)}$ is 20. Using the relationships deduced above

$$R_L = V_{CC}/I_{C(sat)} = 10\text{ V}/5\text{ mA} = 2\text{ k}\Omega$$

Since we have specified a *maximum* value of collector current of 5 mA, it is advisable to select a value of $R_L$ which is the next preferred value above 2 kΩ, that is, $R_L = 2.2$ kΩ.

The preferred range of 10 per cent tolerance resistors are decimal multiples and sub-multiples of the following range: 10, 12, 15, 18, 22, 27, 33, 39, 43, 47, 56, 68, and 82.

If we can assume that the logic '1' voltage is always 10 V, then we may use a value for $R_1$ of

$$R_1 = h_{FE(sat)} R_L = 20 \times 2.2 = 44\text{ k}\Omega$$

If the output of the gate is loaded so that the logic '1' level falls to $V_{CC}/2$, then $R_1$ will need to have a value of 22 kΩ if the transistor is to be saturated. Furthermore, if we make allowances for all the factors mentioned above, we really need to select a value for $R_1$ of about

$$R_1 = 0.3 \times 44 = 13.2 \text{ k}\Omega$$

Using the next *lower* preferred value, we would select

$$R_1 = 12 \text{ k}\Omega$$

## 5.6 THERMAL CONSIDERATIONS

Temperature changes affect the fan-in, the fan-out, the switching speed, and the logic levels of the gates. Gates used in industrial and commercial applications are designed to operate within their specified limits over a temperature range 0 to 75 °C. Military systems and certain industrial systems which need a wider operating temperature range, use families of logic gates which have an operating temperature range from −55 to 125 °C.

## 5.7 NOISE IMMUNITY

The noise immunity of a logic gate is the degree to which it can withstand variations in logic levels at the input without causing the output state of the gate to change significantly.

The d.c. noise margins can be defined in terms of the transfer characteristic of the gate, an example being shown in figure 5.7. In the characteristic shown, the HIGH output voltage is 3 V and the LOW output voltage is 0.2 V. Point A on the characteristic corresponds to the operation of the circuit when the input is energised by a LOW signal (0.2 V) and point B to the operating state when the input is HIGH (3 V). In the case considered, the transition from one logic level to the other occurs between input voltages of 0.9 V and 1.1 V. A positive-going 'noise' signal of (0.9 − 0.2) = 0.7 V can be superimposed on the minimum logic '0' input state without the circuit malfunctioning. Also, a negative-going 'noise' signal of (3 − 1.1) = 1.9 V can be superimposed on the maximum logic '1' input state without serious effect. However, if the driving gate is loaded so that the logic '1' signal applied to the input is only 1.5 V, then the noise margin for the HIGH input state is only 0.4 V.

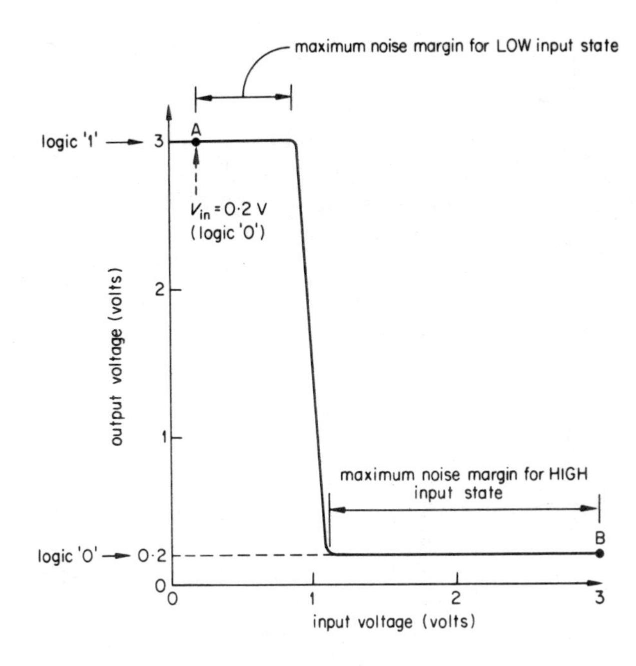


Figure 5.7 A typical transfer characteristic for an INVERTING gate

Similarly, the LOW input noise margin is affected by the maximum allowable '0' voltage level. If, for example, this is 0.4 V, then the LOW input noise margin is only (0.9 − 0.4) = 0.5 V.

## 5.8 CONNECTION OF UNUSED INPUT LINES

It is not good practice to leave unused input lines disconnected, since electronic 'noise' voltages can be induced in them. This may cause the system to behave in an unpredictable fashion. *Thus all unused input lines should be connected to some point which renders them harmless.*

The connection used depends on the type of gate. In all types of gate, unused input lines may be connected to used inputs as shown in figure 5.8a. This has the merit of simplicity, but adds a 'unit' load to the driving gate for each connected input line and reduces the fan-out of the driving gate by unity.

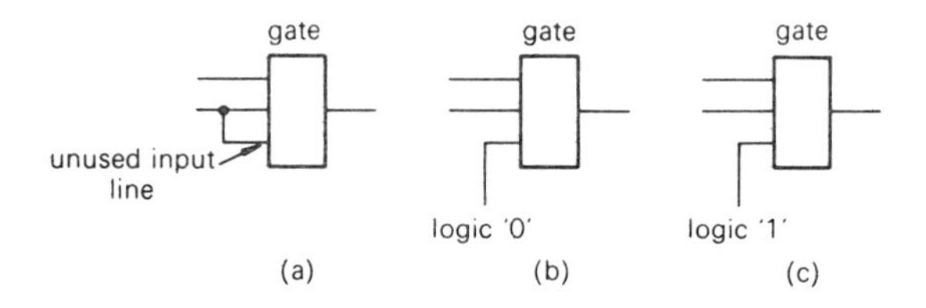


Figure 5.8 (a) In all types of gates, unused inputs may be connected to used inputs. Connection (b) can be used for OR and NOR gates, and (c) may be used for AND and NAND gates

In the case of OR and NOR gates, a logic '0' at any input does not affect the operation of the gate; in these gates any unused input lines may be connected to a logic '0' line as shown in figure 5.8b. In the case of AND and NAND gates, a logic '1' at any input has no effect on the operation of the gate; in these gates, unused input lines may be connected to a logic '1' line as illustrated in figure 5.8c.

## 5.9 TIME DELAYS IN A BIPOLAR TRANSISTOR SWITCH

The speed with which a signal can be propagated through an electronic gate depends both on the number of time delays involved and also on their magnitude. Typical switching waveforms associated with a NOT gate are shown in figure 5.9.

For a short period of time after the base drive has been applied, the transistor remains in the *off* state and the collector current is zero. This period of time is known as the *delay time*, $t_d$, and is the time needed for the base current to propagate through the base

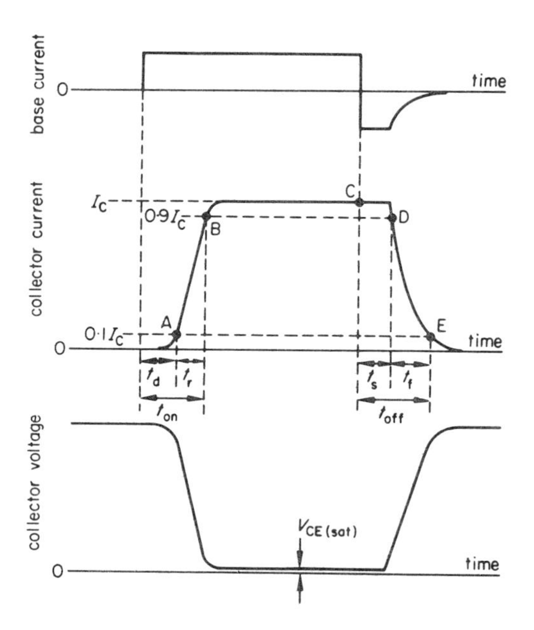


Figure 5.9 Switching waveforms of an INVERTING gate

region in order to begin to bring the transistor into a conducting state. The *rise time*, $t_r$, is the time taken for the collector current to rise from 10 per cent to 90 per cent of its maximum value. During this interval of time and with a resistive load, the collector voltage falls from 90 per cent to 10 per cent of its maximum value. The rise time is determined by several factors including the high-frequency response of the transistor, its current gain, and the magnitude of the base drive. Transistors used in switching applications should have as high a cut-off frequency as possible.

When the base voltage is reduced to zero (point C in figure 5.9), the collector current remains constant for a period of time known as the *storage time*, $t_s$, which is the time required to remove excess base charge. The collector current falls from 90 per cent to 10 per cent of its maximum value in a time interval known as the *fall time*, $t_f$, and during this interval the collector voltage rises from 10 per cent to 90 per cent of its maximum value.

The total time required to turn the transistor *on* is known as the *turn-on time*, $t_{on}$, and the total time taken to turn it *off* is the *turn-off time*, $t_{off}$, where

$$t_{on} = t_d + t_r$$

and

$$t_{off} = t_s + t_f$$

Typical values for a silicon switching transistor are $t_{on} = 12$ ns, $t_{off} = 15$ ns, with $t_s = 10$ ns.

An important factor in specifying the switching performance of gates is the time taken for a signal to propagate through the gate from its input to its output, and is known as the *propagation delay*, $t_{pd}$, and is defined in terms of the waveforms in figure 5.10b. The waveforms relate to the inverting gate in figure 5.10a. The

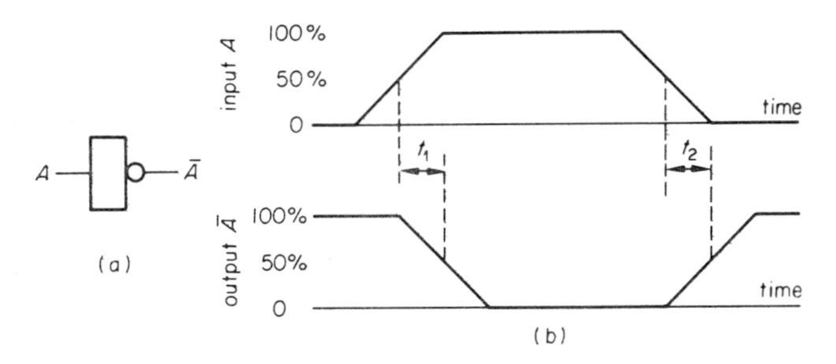


Figure 5.10 Propagation delay

propagation delay is specified at the 50 per cent voltage levels of input and output signals, where

$$t_{pd} = \frac{t_1 + t_2}{2}$$

The propagation delay depends on many factors including the circuit design, the supply voltage and the power consumption, and may have a value from about 1 ns to a fraction of a microsecond. Values for typical logic families are given later in this chapter.

## 5.10 METHODS OF IMPROVING THE SWITCHING SPEED

Two methods used to reduce the switching time of gates are illustrated in figure 5.11. The first of these methods is to shunt the input resistor $R$ with a *speed-up capacitor C*. This allows the total input voltage transition to be momentarily applied to the base of the transistor. Thus, when signal $A$ changes from zero to a positive voltage, the capacitor charging current causes a momentary rush of current to be applied to the transistor base, so reducing the delay time. When signal $A$ is reduced to zero, the negative-going voltage transition is transmitted to the transistor base. This has the effect of reducing the storage time.

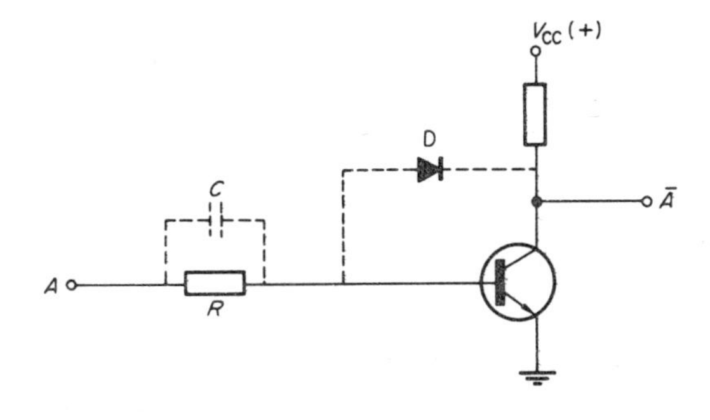


Figure 5.11 Methods of improving the switching speed of an INVERTING gate

Unfortunately, the speed-up capacitor allows noise signals to be transmitted to the transistor base, so that it reduces the noise margin to some extent. In many cases, in order to obtain an improved switching performance, it is advisable to use a transistor with a higher cut-off frequency rather than to use a speed-up capacitor.

As we have seen, a factor which limits the switching speed is the storage time of the transistor. The storage time can be reduced by preventing the transistor from being driven too hard into saturation, that is, by limiting the base current to a value which is just sufficient to saturate the transistor. One method of achieving this solution is by the use of the *clamping diode* D. The onset of saturation occurs when the collector voltage falls below the base voltage; when this occurs diode D becomes forward biased and the excess base current is diverted via diode D and the transistor to earth. Since the diode only carries the excess base current, the storage time of the diode is much shorter than that of an overdriven transistor.

Ideally, the clamping diode should have zero storage time which is theoretically obtained in Schottky diodes (see also section 5.14).

## 5.11 ACTIVE COLLECTOR LOADS

An *active* circuit element is one which provides voltage gain or current gain, examples of which are bipolar transistors and field-effect transistors. A *passive* circuit element is one which does not exhibit the property of gain, examples of which are resistors, capacitors and inductors.

In the NOT circuit described earlier, a passive resistor (often referred to as a *pull-up resistor*) was used as the collector load of the transistor. In order to change the value of the current flowing through the resistor, it is necessary to either charge or discharge the stray capacitance associated with that resistor. If the resistor is replaced by an *active load* then, as a result of the gain of the active device, the rate at which the output voltage can change is accelerated. One form of active load is illustrated in figure 5.12. Here the active load comprises transistor TR2 (the *pull-up transistor*), diode D, and resistor R. The circuit operates as follows. When $X = 1$, transistor TR1 is saturated and the output voltage is LOW, that is, logic '0'. The presence of diode D in the circuit causes the emitter voltage of TR2 to be higher than its base voltage, so that TR2 is cut-off. When $X = 0$, transistor TR1 is *off*, so that the output voltage depends on the operation of the active load. In this condition, current flows into the base of TR2 via resistor $R$, causing TR2 to saturate. Thus the voltage drop between the supply line and the output terminal is the $V_{CE(sat)}$ of TR2, that is, the output voltage is very nearly equal to $V_{CC}$. Hence, we see that an input of logic '0' gives a logic '1' output whose value is very nearly equal to $V_{CC}$.

Diode D fulfills the important function of preventing both TR1 and TR2 from being switched *on* simultaneously. If it were possible for the two transistors to be turned *on* together, even for a few nanoseconds, the resulting current spike would generate a great deal of electronic noise and would also cause excessive power dissipation in the circuit.

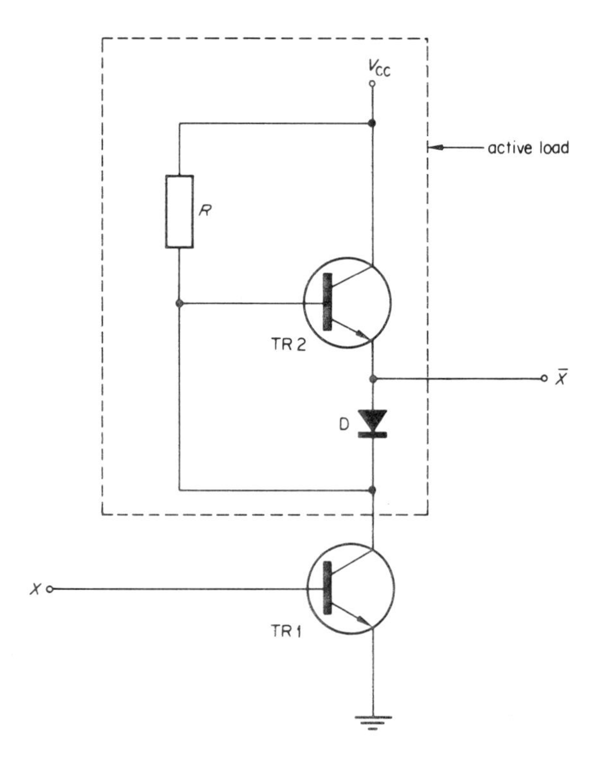


Figure 5.12 One form of active collector load

## 5.12 THE WIRED-OR FUNCTION

It is convenient in many cases to generate logical functions by using what is known as the wired-OR connection of logic gates, in which the output terminals of the gates are connected together. The general arrangement is shown in figure 5.13a, with the symbolic

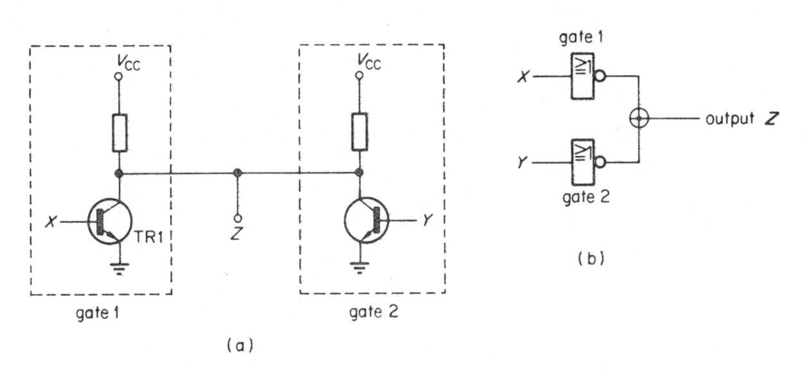


Figure 5.13 The wired-OR connection

representation in figure 5.13b. Although NOR-type invertors are shown in figure 5.13b, this connection can, with the exception given below, also be used with NAND gates.

In figure 5.13a, when the signal $X = 1$, TR1 is saturated so that $Z = 0$ irrespective of the logical value of $Y$. Similarly, when $Y = 1$, $Z = 0$ irrespective of the value of $X$. The truth table for figure 5.13a is given in table 5.2.

**Table 5.2**

| $X$ | $Y$ | $Z$ |
|---|---|---|
| 0 | 0 | 1 |
| 0 | 1 | 0 |
| 1 | 0 | 0 |
| 1 | 1 | 0 |

Thus $Z = 1$ when $X = 0$ and $Y = 0$, that is

$$Z = \overline{X} \,.\, \overline{Y}$$

Hence the logical output from the circuit is given by the relationship

$Z =$ NOT (the function at the base of TR1) AND NOT (the function at the base of TR2)

In the general case, the functions generated at the bases of the transistors are fairly complex, so that the overall function is usually more complex than that given above. For example, if $X = A \,.\, B$ and $Y = C \,.\, D$, then

$$Z = (\overline{A \,.\, B}) \,.\, (\overline{C \,.\, D}) = (\overline{A} + \overline{B}) \,.\, (\overline{C} + \overline{D})$$

An important point to note with this connection is that both gates use passive pull-up resistors. If active pull-up transistors are used then, when $X = 1$ and $Y = 0$ or when $X = 0$ and $Y = 1$, a continuous short-circuit is applied to the power supply via the wired-OR link. *It is general practice, therefore, to use only gates with resistive loads in wired-OR networks.*

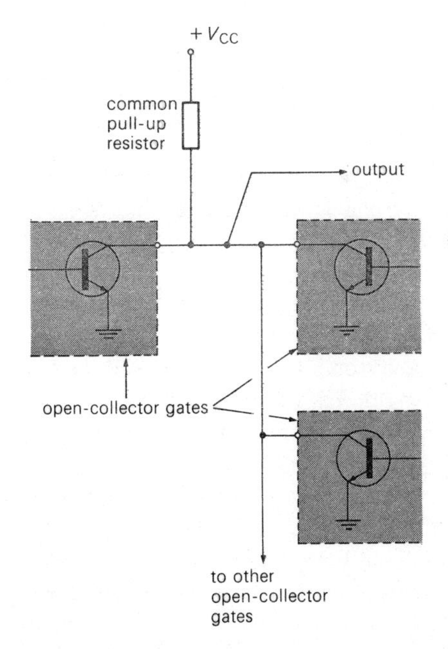


Figure 5.14 Using open-collector gates in the wired-OR configuration

Certain types of TTL gate (see section 5.14) have what is known as an *open-collector output*, in which there is an open-circuit between the collector of the transistor and the supply (see figure 5.14). This allows the user to wire all the outputs to the supply via a common pull-up resistor.

## 5.13 DIODE–TRANSISTOR LOGIC (DTL)

A basic form of DTL NAND gate is illustrated in figure 5.15, and consists of a DRL AND section followed by an invertor. The function generated at point Y is the AND function of inputs $A$ and $B$, and is inverted by the transistor output stage.

When either input $A$ or input $B$ is logic '0', the circuit designer must ensure that the current flowing in $R_1$ chooses to flow through the appropriate input diode ($D_A$ or $D_B$), rather than through D1 and D2. If this were not the case, the current flowing into the base of the transistor would not fall to zero when either input is zero. A preferential bias is given to the flow of current for a logic '0' input by including the two *voltage offset diodes* D1 and D2 in series with the base. The forward biased potential difference across these diodes, together with $V_{BE(sat)}$ of TR, exceeds the sum of the maximum logic '0' level and the forward potential difference in either $D_A$ or $D_B$.

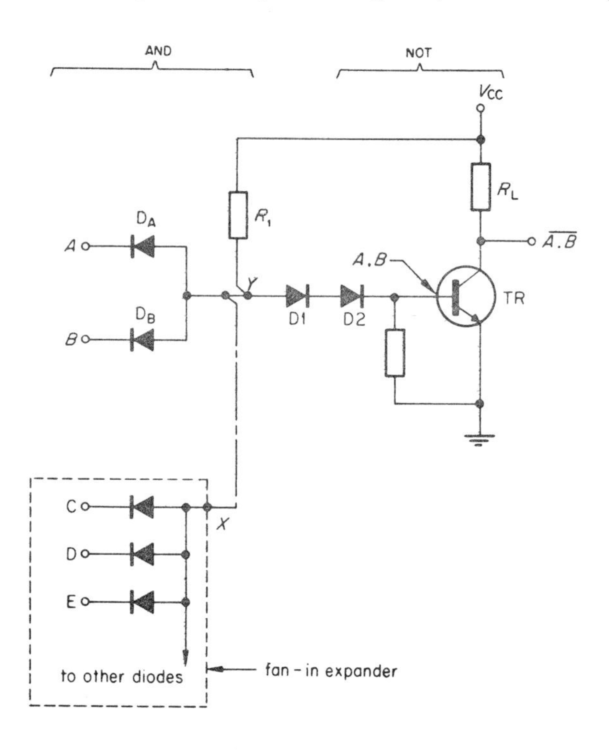


Figure 5.15 A DTL NAND gate

DTL gates of the type shown are *current-sinking gates* within the terms of the definition in section 5.3. The driving transistor is required to 'sink' the current flowing in $R_1$ in each of the driven gates.

The fan-in of the circuit in figure 5.15 can be increased by connecting the diode *fan-in expander* shown in the inset to the figure, points X and Y being linked so that the logic function generated at point Y is $A.B.C.D.E....$

Since the circuit in figure 5.15 employs a passive pull-up resistor $R_L$, it can be used in conjunction with other identical gates in the wired-OR configuration. Certain types of DTL NAND gates employ active pull-up loads similar to the one shown in figure 5.12. The latter gates should not be used in wired-OR networks for the reasons given in section 5.11.

For the circuit shown in figure 5.15, the following values are typical.

| | |
|---|---|
| supply voltage | 6 V |
| typical noise immunity | 1.2 V |
| propagation delay | 30 ns |
| power per gate | 11 mW |
| logic '1' | 4 V (minimum) |
| logic '0' | 0.4 V (maximum) |
| fan-out | 8 |
| fan-in | 14 |

A modified version of the NAND gate which is better suited to integrated-circuit production techniques is shown in figure 5.16. This circuit provides improved fan capability together with a smaller propagation delay when compared with figure 5.15. The improved performance is brought about by replacing D1 in figure 5.15 by transistor TR1 in figure 5.16. In figure 5.16, the driving gate has merely to 'sink' the base current of TR1, rather than all the current flowing in $R_1$.

In an industrial environment it is desirable to have the highest possible value of noise immunity. One version of DTL with a high noise immunity includes the modification shown in the inset in figure 5.16. Diode D2 in the figure is removed, and the circuit shown in the inset is connected to the points marked L, M, N. The Zener diode ZD provides the improved noise immunity, which has a value of approximately $(V_z + 0.7)$ $V$, where $V_z$ is the breakdown voltage of ZD. Zener diodes have a relatively large parasitic capacitance and diode D is included to provide a discharge path for the charge held by this capacitance when the input signal changes to the '0' state.

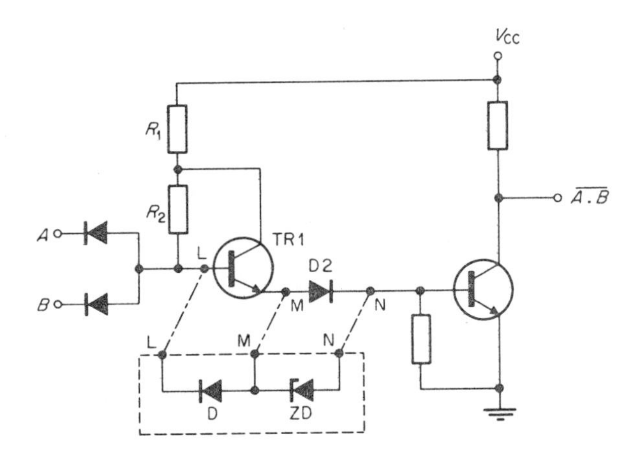


Figure 5.16 An integrated circuit DTL NAND gate

## 5.14 TRANSISTOR–TRANSISTOR LOGIC (TTL)

The development of TTL represented a break from conventional designs and was made possible by developments in monolithic IC production techniques. One basic form of TTL gate, a NAND gate, is shown in figure 5.17, and uses a *multi-emitter transistor* TR1 in the input circuit. Transistor TR2 is used in a phase-splitting amplifier, and provides complementary logical signals at its emitter and collector. Transistor TR4 acts as an active load for transistor TR3, the two being energised by the complementary outputs from TR2. The circuit operation is described below.

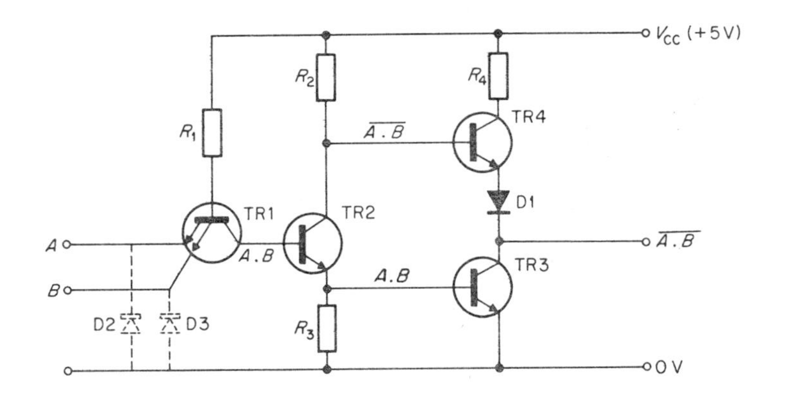


Figure 5.17 A TTL NAND gate

When any input line is at the logic '0' level, the current flowing in $R_1$ is diverted to that line. The flow of base–emitter current causes TR1 to saturate, thereby causing TR1 collector voltage to fall to the logic '0' level; that is, the collector signal of TR1 is the AND function of the inputs. This LOW voltage is applied to the base of TR2 and causes it to be cut-off, so that its emitter voltage is LOW and its collector voltage is HIGH. In turn, this results in TR3 being turned *off* and TR4 being *on*, connecting the output line to the supply voltage. That is, a logic '0' on any input causes the output to be logic '1'.

When all inputs are at logic '1', the current flowing in $R_1$ is diverted through the collector region of TR1 into the base of TR2, resulting in TR2 saturating. This immediately causes the emitter voltage of TR2 to rise and its collector voltage to fall. The HIGH voltage at the emitter of TR2 drives TR3 into saturation, and connects the output terminal to the zero volts line. Also, the reduced collector voltage of TR2 causes TR4 to be cut-off, so that no current flows through TR4.

The reason for the use of diode D becomes apparent when we study the quiescent voltages existing in the circuit in the above operating state. Since, in a silicon transistor, $V_{BE\,(sat)}$ is about 0.7 V and $V_{CE\,(sat)}$ is about 0.2 V and, in a silicon diode, the forward biased potential difference is about 0.7 V, the following voltages exist with all inputs HIGH

| | |
|---|---|
| at the base of TR3 (emitter of TR2) | 0.7 V |
| at the base of TR4 (collector of TR2) | 0.9 V |
| at the collector of TR3 | 0.2 V |

Thus, the potential between the base of TR4 and the collector of TR3 is only about 0.7 V, which is insufficient to cause TR4 and D1 to conduct. Had D1 not been included, this voltage would be sufficient to maintain TR4 in a conducting state. The output circuit shown in figure 5.17 is sometimes known as a *totem-pole* circuit because of its shape.

When the output voltage is caused to change from one level to the other, both TR3 and TR4 may conduct simultaneously for a very short interval of time, resulting in a surge of current being drawn from the supply. Resistor $R_4$ is included to limit the value of this surge.

A typical transfer characteristic for a TTL gate is shown in figure 5.18. Using a 5 V supply, the logic '1' output is constant at 3.3 V until the input voltage reaches 0.7 V (point A). At this point TR2 in figure 5.17 begins to conduct, but TR3 has not yet begun to carry current. Between points A and B on the characteristic, TR2 operates as part of a linear amplifier having a gain of $-R_2/R_3$ (that is, about $-1.6$ for the standard gate). When the input voltage reaches 1.4 V, at point B, TR3 begins to conduct, and the output voltage rapidly falls to $V_{CE\,(sat)}$ which is about 0.2 V.

The very fast rise and fall-times associated with TTL gates (about 1.5 ns/V) sometimes bring in their wake oscillations of voltage on the output line. Under certain circumstances, a voltage undershoot of greater than $-2$ V may be applied to the input of a driven gate. One solution sometimes adopted to overcome this problem is the use of the *diodes* D2 and D3 shown in figure 5.17. Reflections can occur along any line when the rise or fall time of the

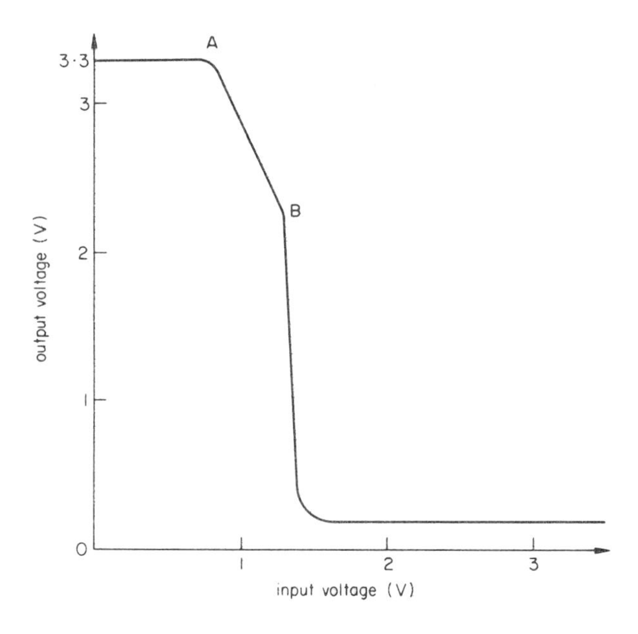


Figure 5.18 Typical transfer characteristic of a TTL gate

propagated wave is comparable with the time taken for a signal to travel the length of the line. An electrical signal travels at the rate of about 0.3 m/ns hence, since the rise-time of a TTL gate is about 5 ns, a line is regarded as being electrically long in a TTL system when its length is about 1.5 m.

In TTL networks, in order to avoid the possibility of spurious switching from the pick-up voltages on unused input lines, spare input lines should either be connected to the +5 V line or connected in parallel with used inputs.

The basic TTL gate in figure 5.17 cannot be used in wired-OR networks because of the active load TR4. Special types of *open-collector* TTL gates (in which $R_4$, TR4 and D1 are omitted) are manufactured and can be used in conjunction with an external resistor in the wired-OR configuration (see section 5.12).

A stabilised power supply is essential for use with TTL systems, and the supply voltage is specified in the range 5 V $\pm$ 0.25 V. These voltage limits are sometimes difficult to maintain due to the fact that both transistors in the totem-pole output circuit are *on* simultaneously for a short period of time when the output is changing from '0' to '1' and vice versa. Consequently, it is necessary to distribute a number of capacitors throughout the system (preferably being connected to the supply terminals of the IC) to provide the momentary current surge when switching occurs.

The TTL family has six basic branches known respectively as *standard TTL, high-power TTL, low-power TTL, Schottky TTL, lower-power Schottky TTL and three-state TTL* (or *tri-state TTL*).

The essential difference between the first three types is the value chosen for the resistors in the circuit. In high-power gates the resistor values are lower than in standard gates, leading to a higher current consumption and a higher power dissipation. In low-power gates, the resistor values are higher than in the standard version, resulting in a lower current consumption. The lower resistance values in the high power gates are associated with a low time constant, so that the propagation delay of high-power gates is reduced when compared both with standard and low-power gates. By a similar argument it can be seen that the propagation delay of a low-power gate is higher than that of a standard TTL gate. Typical performance parameters of a standard TTL NAND gate are

| | |
|---|---|
| supply voltage | 5 V |
| logic '1' | 3.3 V |
| logic '0' | 0.2 V |
| propagation delay | 10 ns |
| fan-out | 10 |

It has been shown in section 5.10 that the switching speed of an inverting circuit is increased by the use of a clamping diode which prevents the transistor from being driven too far into saturation. The use of Schottky diode clamped TTL (*Schottky TTL*) allows the propagation delay of a single gate to be reduced to about 3 ns. *Low-power Schottky TTL* uses higher values of resistance than Schottky TTL, resulting in a gate which consumes a low power whilst giving a propagation delay of about the same order as that of standard TTL.

The basis of a *three-state gate* is shown in figure 5.19a. Switch S is part of the electronics of the gate and is controlled by the logic

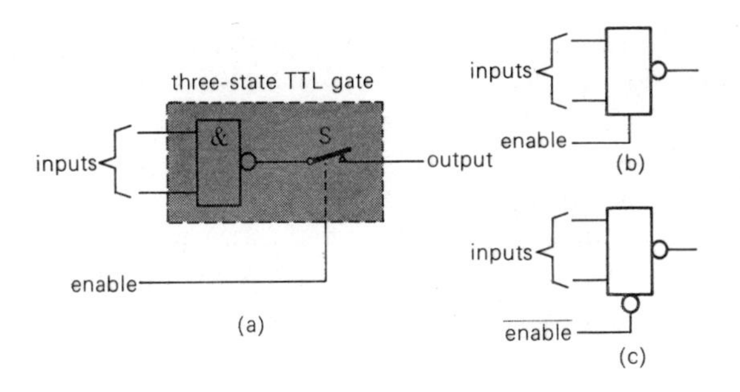


Figure 5.19 (a) The basis of three-state TTL, (b) a 'high' enable gate and (c) a 'low' enable gate

signal applied to the *enable* input of the gate. When the gate is 'enabled' it functions normally and the logic function of the input signals appears at the output. When the gate is 'disabled', switch S is open and the logic function generated by the gate is isolated from or is disconnected from the output line. Switch S is effectively 'opened' by reducing the base current of both output transistors (TR3 and TR4 in figure 5.17) to zero. The circuit of a three-state gate does, in fact, differ from that in figure 5.17, but the output circuit is a version of the totem-pole circuit.

Gates are available which can be enabled either by a logic '1', that is, a 'high' enable (figure 5.19b), or by a logic '0', that is, a 'low' enable (figure 5.19c). In the latter case, the signal which operates switch S is sometimes described as an $\overline{\text{enable}}$ signal.

An advantage offered by three-state logic over other branches of TTL is that it allows complex networks of gates to operate with a common set of busbars along which information can be transmitted. A simple example of this type of network is illustrated in figure 5.20. Signal *A* is used to 'enable' gates G1 and G1a, while signal *B* enables G2 and G2a and signal *C* enables G3 and G3a. It is arranged that when a logic '1' is applied to *A*, then '0's appear on *B* and *C*; when a '1' is applied to *B*, then '0's appear on *A* and *C*, and so on. Thus when gates G1 and G1a are simultaneously enabled, the signals from G2 and G3 are disconnected from the common busbar; at the same time, information cannot be transmitted through G2a and G3a. Under this operating condition, the signal

$\overline{U.V}$ at the output of G1 is transmitted along the common busbar and appears at the output of G1a. When a logic '1' is applied to line $B$ it enables G2 and G2a, and the signal $\overline{W.X}$ appears at the output of G2a.

The system illustrated in figure 5.20 only has the ability to transmit information from left to right. In a practical system the circuit would be modified to allow data to be transmitted in both directions; this technique is widely adopted in microprocessors and computers.

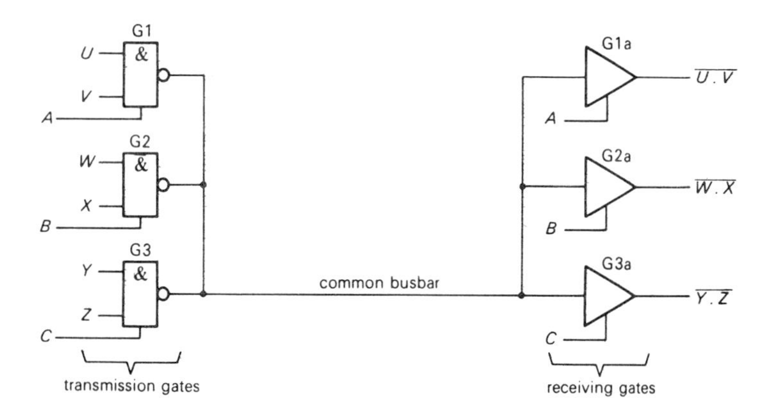


Figure 5.20 The use of three-state gates

## 5.15 EMITTER-COUPLED LOGIC (ECL)

The time delays associated with charge storage in the saturated logic gates described earlier can be eliminated if the transistors are not allowed to saturate. The ECL family of gates, with a typical OR/NOR gate being shown in figure 5.21, were developed with this purpose in mind. The propagation delay of this family of gates is typically 2 ns, and the fan-out is typically 25. Its drawbacks include a high power dissipation per gate (about 60 mW) and high sensitivity to temperature changes.

The basis of the circuit is a non-saturated emitter-coupled amplifier which consists of two sections, the right-hand section containing TR3 and the left-hand section containing TR1 and

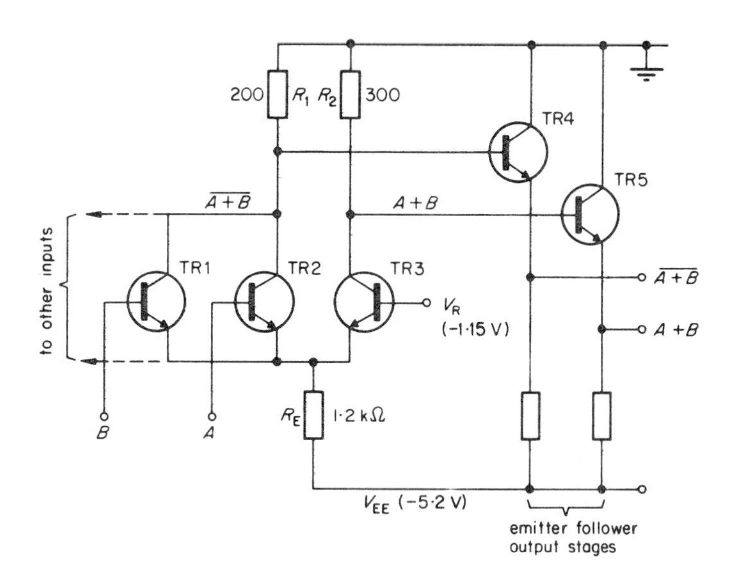


Figure 5.21 An ECL OR/NOR gate

TR2. The value of $R_E$ is large compared with both $R_1$ and $R_2$, so that the current flowing through the emitter-coupled amplifier is largely fixed by the value of $R_E$. The base of TR3 is supplied by a voltage reference source $V_R$ of $-1.15$ V. The circuit operates with positive logic, logic '1' corresponding to an output of $-0.75$ V, and logic '0' to $-1.55$ V, giving a voltage swing of 0.8 V between the two levels. Thus, when $A = B = 0$, the current in $R_1$ is at its minimum value and in $R_2$ is at its maximum value. When either or both of the inputs are logic '1', the current in $R_1$ is increased and that in $R_2$ is reduced. Since TR1 and TR2 are connected in a wired-OR configuration, the output at the collector of TR1 is $\bar{A}.\bar{B}$ which alternatively is $\overline{A+B}$, that is, the NOR function of the inputs. As a result of the action described above, we see that the logical function at the collector of TR3 is $A+B$.

The emitter follower stages containing TR4 and TR5 fulfil two functions. Firstly, they provide a low-impedance output which results in a large fan-out. Secondly, the $V_{BE}$ drop of TR4 and TR5 restores the output voltages to the correct levels for driving other stages.

The following values are typical of an ECL OR/NOR gate.

| | |
|---|---|
| supply voltage ($V_{EE}$) | −5.2 V |
| logic '1' | −0.75 V |
| logic '0' | −1.55 V |
| fan-out | 25 |
| propagation delay | 2 ns |

Other names used to describe ECL are *current-mode logic* (CML), *emitter–emitter coupled logic* ($E^2CL$), and *emitter-coupled transistor logic* (ECTL).

## 5.16 MOS LOGIC GATES

A *p*-MOS 2-input NOR gate using two MOSFETs in the wired-OR connection is shown in figure 5.22a and a 2-input NAND gate

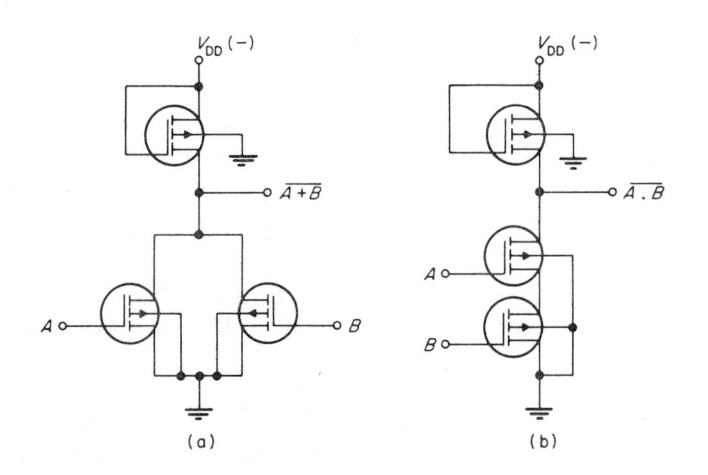


Figure 5.22 (a) A NOR gate and (b) a NAND gate using *p*-channel MOS devices

using series connected FETs is illustrated in figure 5.22b. These circuits operate in negative logic. With a drain supply voltage $V_{DD}$ of −20 V, the following values are typical of *p*-MOS gates.

| | |
|---|---|
| propagation delay | 100 ns |
| power per gate | 7 mW for logic '0' output<br>zero for logic '1' output |
| logic '1' | −11 V (minimum) |
| logic '0' | −3 V (maximum) |

The circuit in figure 5.23a is that of a 2-input positive logic CMOS NOR gate, and that in figure 5.23b is for a positive logic 2-input CMOS NAND gate. The use of *n*-channel devices reduces the propagation delay and allows the maximum operating frequency of CMOS devices to be doubled when compared with that of *p*-MOS devices.

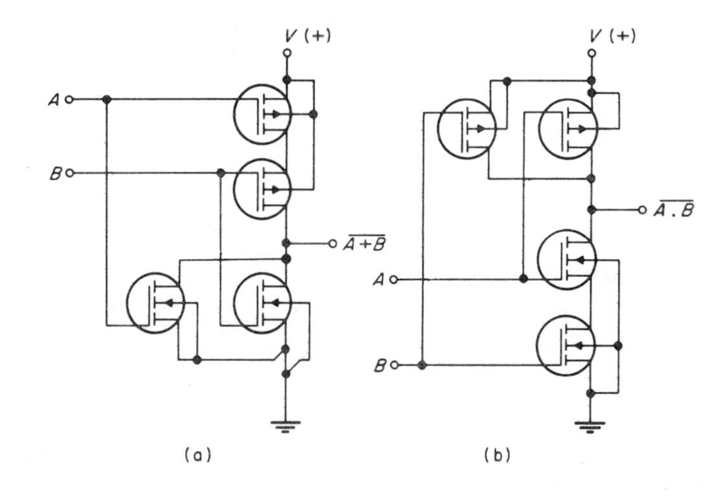


Figure 5.23 (a) A NOR gate and (b) a NAND gate using complementary MOS (CMOS) logic

CMOS devices can operate over a very wide range of supply voltages, 3–15 V being typical. This allows CMOS gates to be operated from the same power source as TTL and DTL systems. The noise immunity of CMOS is about 0.45 $V_{DD}$. The following figures apply to one family of CMOS gates when operating on a supply of 5 V.

| | |
|---|---|
| noise immunity | 2.25 V |
| propagation delay | 40 ns |
| power per gate | 5 nW |
| logic '1' | 4.99 V (minimum) |
| logic '0' | 0.01 V (maximum) |

The very low quiescent power consumption arises from the fact that when the *p*-channel devices are *on* then the *n*-channel devices are *off*, and vice versa. Also the fan-out is large since the current consumption per driven gate is only about 10 pA.

Since the insulating layer between the gate and the conducting channel of a FET is very thin, it is easily damaged by the application of a voltage above 80–100 V. Electrostatic voltages, such as may be generated simply by wearing nylon clothing, are sufficient to rupture the gate insulation. To minimise the possibility of damage from this cause, each input line of a CMOS gate usually incorporates a *gate-oxide protection circuit* similar to that shown in figure 5.24.

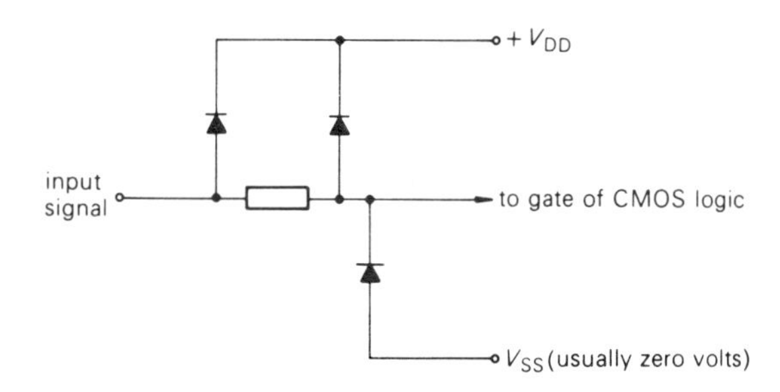


Figure 5.24 Typical gate-input protection circuit for a CMOS gate

## 5.17 CMOS TRANSMISSION GATES (BILATERAL SWITCHES) AND THREE-STATE LOGIC

A CMOS transmission gate or bilateral switch is an electronic switch which allows signals (either digital or analogue) to be transmitted through the gate in either direction (that is, from the input to the output or vice versa).

A simplified equivalent circuit of a transmission gate is shown in figure 5.25a. When the signal applied to the enable/disable line is logic '0', the switch is open and the resistance between the input and output terminals is typically greater than 1000 MΩ. The application of a logic '1' signal to the control line causes the bilateral switch to close, when the resistance between the input and output is typically less than 300 Ω.

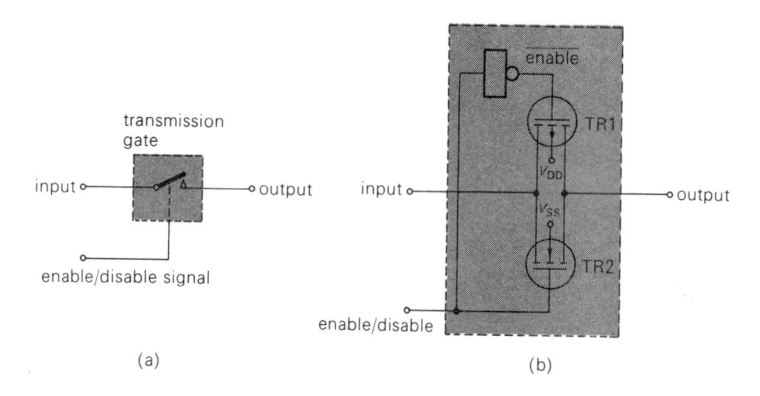


Figure 5.25 (a) Equivalent circuit of a transmission gate or bilateral switch and (b) its basic circuit

The circuit of a basic transmission gate is shown in figure 5.25b. Transistor TR1 is a *p*-channel MOSFET and TR2 is an *n*-channel device; complementary logic signals are applied to the gate of TR1 and TR2 to ensure that either both are *on* simultaneously or that both are *off*.

The ways in which a transmission gate can be used in conjunction with logic circuits are varied, one method of implementing a three-state CMOS gate being shown in figure 5.26. A logic '1' applied to the enable/disable line 'enables' the transmission gate so that the signal from G1 is connected to the 'output' terminal of the three-state gate. A logic '0' on the control line 'disables' the transmission gate, resulting in the signal from G1 being isolated from the 'output' line. This facility allows CMOS three-state gates to be used on a common busbar system of the type described in section 5.14.

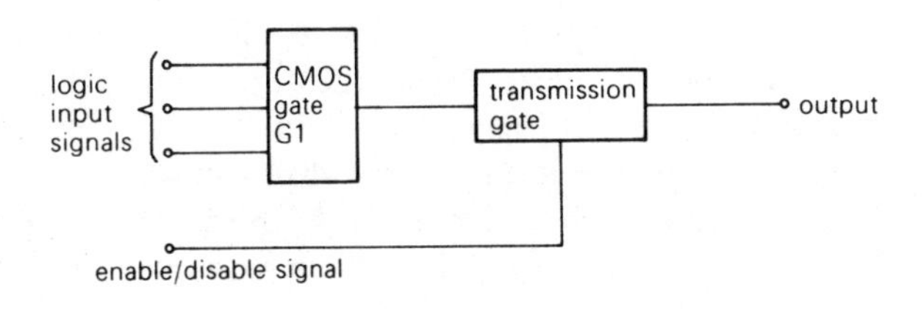


Figure 5.26 Basic three-state CMOS gate

## 5.18 INTEGRATED INJECTION LOGIC ($I^2L$)

$I^2L$ gates (also known as *merged transistor logic*, MTL) are constructed using bipolar transistors (*n–p–n* and *p–n–p* types), which not only provide a very high density of gates on the chip surface, but also give a very short propagation delay to signals passing through them. That is to say, they offer the high packing density of MOS gates combined with the low propagation delay of TTL.

A section through the basic structure is shown in figure 5.27a, and consists of a lateral *p–n–p* transistor TR1, which acts as a constant current source for the base circuit of the multiple-collector *n–p–n* transistor TR2. Transistor TR1 is described as a *current injector*. The logic input signal is applied to the base *B* of TR2, and the output is taken from a collector electrode ($C_1$ or $C_2$). The logic state of the output is switched from one level to another when the injector current *I* is switched either to or from the base of TR2 by signal *B*. The circuit representation of the gate is shown in figure 5.27b.

If the input signal applied to *B* is 'low' (less than about 0.75 V), the injector current is steered away from the base of TR2, so that TR2 is cut off. The logic level at the collector ($C_1$, $C_2$) is then 'high'; that is a 'low' input signal gives a 'high' output signal. When the input signal is 'high' (greater than about 0.75 V), the injector current is steered into the base of TR2. This causes TR2 to saturate so that the output voltage is 'low'.

Using a combination of $I^2L$ elements, it is possible to implement either NAND-type or NOR-type logic by interconnecting the outputs and inputs of $I^2L$ gates within an integrated circuit.



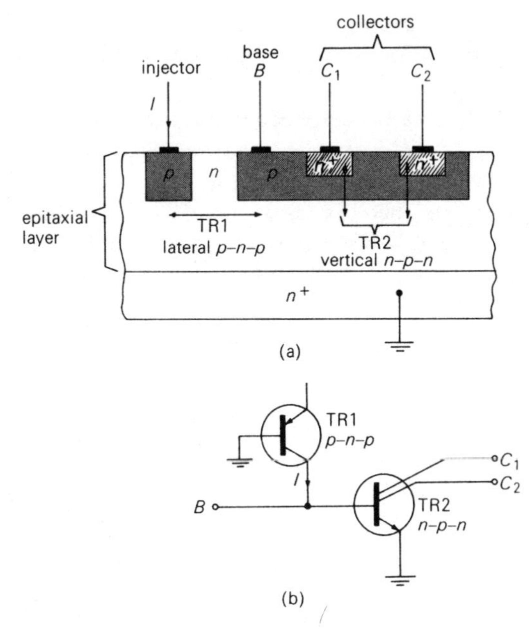


Figure 5.27 (a) Section through an $I^2L$ element, (b) its circuit representation

$I^2L$ gates will operate from a power source of 850 mV or greater and the injector current can have any value in the range 1 nA to a few mA. It is possible to increase the switching speed simply by increasing the injection current. There is therefore a trade-off at the design stage between power consumption and switching speed. However, a limit exists to the power consumption beyond which the switching speed does not increase.

## 5.19 MONOLITHIC INTEGRATED CIRCUIT CONSTRUCTION

Integrated circuits used in electronic logic systems are usually constructed from silicon in what is known as *monolithic planar* form. That is, they are manufactured in a single slice of silicon in a 'flat' or plane form, and do not normally require the use of external components. For example, the integrated circuit shown in cross-section in figure 5.28b is a monolithic form of the bipolar circuit in figure 5.28a.

In the following, a simplified version of the construction of figure 5.28 is given. First, the silicon is refined into a cylindrical ingot, which is then cut up into a number of *slices*, each about 200 μm thick. For the purpose of comparison, the thickness of the paper on which this book is printed is about 100 μm. The slice forms the *substrate* of the IC, and is of *p*-type material.

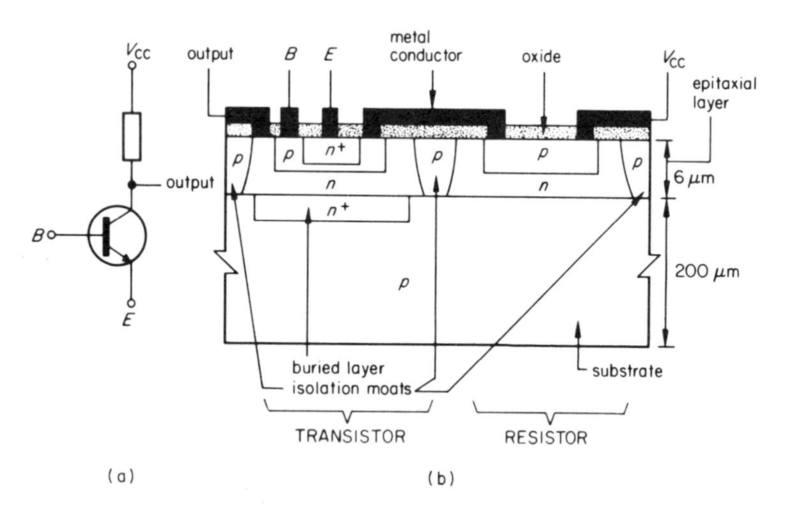


Figure 5.28 A monolithic integrated circuit

An $n^+$ layer is then *diffused* in a special furnace into the substrate, and this layer is eventually 'buried' under an *n*-type *epitaxial layer* (meaning 'arranged upon'). The $n^+$ symbol simply implies that the impurity doping of the $n^+$ region is higher than is normal in an *n*-type material: it results in the $n^+$ region having an increased conductivity relative to an *n*-region. The high conductivity buried layer is introduced to reduce the saturation voltage $V_{CE(sat)}$ of the transistor.

An *n*-type epitaxial layer is then *grown* upon the substrate, and it is in this layer that the complete IC is constructed. The next step is the diffusion of the *p*-type regions which form the isolation moats between the circuit components. In bipolar ICs, fairly complex isolation techniques need to be adopted to ensure isolation between components. A further *p*-type diffusion follows to form the base of the transistor and the body of the pull-up resistor. A final $n^+$ diffusion results in the formation of the transistor emitter. The whole surface is then covered by a thin layer of glass insulation (silicon dioxide), through which 'windows' are etched to allow aluminium connections to be made to the electrodes.

Figure 5.29 shows a simplified cross-sectional diagram of a *p*-MOS NAND gate, corresponding to the circuit in figure 5.22b. The diagram shows the relative simplicity of MOS IC construction when compared with bipolar construction. In figure 5.29 the channel length of the pinch-effect transistor is about three times that of the transistors. Also, the channel width of the resistor is about one-quarter that of the transistors.

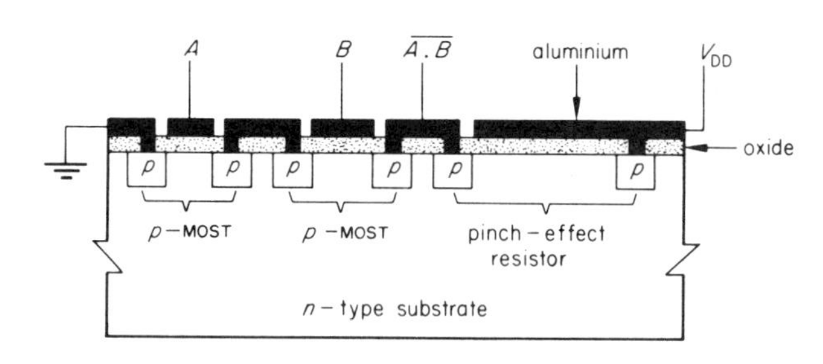


Figure 5.29 A MOS integrated circuit

In the construction of CMOS ICs it is necessary to provide isolation between the *p*-channel and *n*-channel devices, thereby reducing the component density in the surface of the IC. One method of achieving isolation is to diffuse a 'tub' of *p*-type material into the substrate, into which the *n*-channel devices are introduced.

## 5.20 LSI, MSI, AND SSI

The terms *large-scale integrated (LSI) circuit*, *medium-scale integrated (MSI) circuit*, and *small-scale integrated (SSI) circuit* are in everyday use, and refer to the number of gates contained in a single IC package. The terms are not always precisely defined, and the following are those in common usage

| | |
|---|---|
| LSI | contain more than about 100 gates |
| MSI | contain between about 10 and 100 gates |
| SSI | contain up to about 10 gates |

An LSI chip contains a complete logic system, a microprocessor chip being a typical example. An MSI chip contains an interconnected series of elements, examples being shift registers and adders. SSI chips contain a limited number of gates, the 7400 TTL NAND package described below is an example of this kind.

## 5.21 IC PACKAGING

Monolithic ICs are manufactured in three basic types of package (or *packs*), which are

(1) T05 canisters (or *cans*)
(2) flatpacks
(3) plastic encapsulated dual-in-line packs (DIP)

Outline diagrams of the three types are shown in figure 5.30, types 1, 2, and 3 above corresponding to diagrams a, b and c, respectively, in figure 5.30.

Both flatpacks and T05 cans are hermetically sealed, and can be used over a temperature range of −55 to 125 °C and are generally used where space and weight are at a premium. The plastic DIP is cheap to manufacture and is used in the great majority of industrial and commercial applications. With the normal range of devices, DIPs can operate over the temperature range 0 to 70 °C, and some types can be used over the wider range given above.

The most popular arrangement is the 14-pin DIP, having seven connections or pins per side, the pins being 2.5 mm (0.1 in.) apart. Of these, one is required for the power supply and one for the earth line, leaving twelve connections for the input and output of data. A widely used 14-pin DIP is the 7400N (or FJH131) TTL quadruple (or *quad*) 2-input NAND pack, the connections to which are shown in figure 5.31.

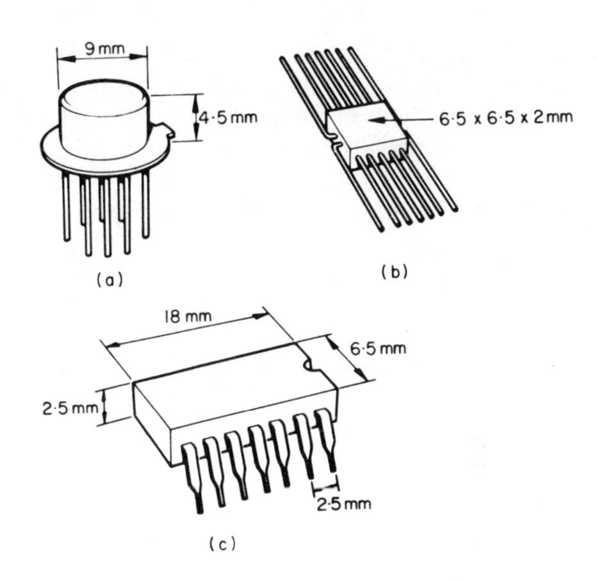


Figure 5.30 Methods of packaging integrated circuits

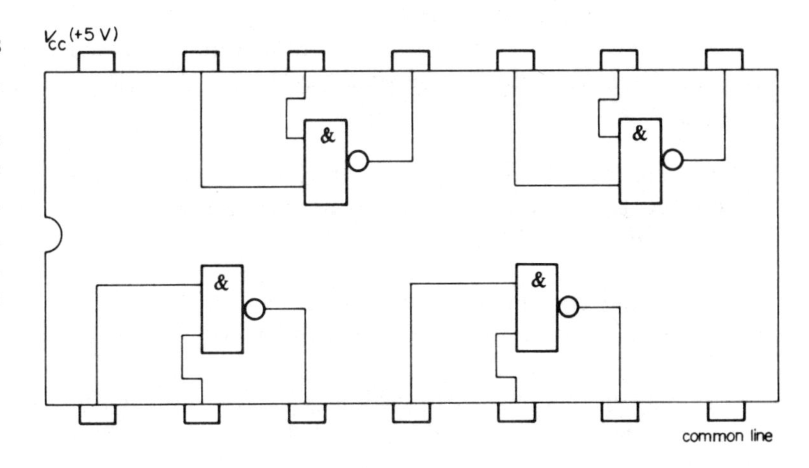


Figure 5.31 A quadruple (quad) 2-input NAND integrated-circuit package

## PROBLEMS

**5.1** Discuss the limitations of diode–resistor logic with regard to fan-in, fan-out and switching speed.

**5.2** If the logic signals applied to inputs $A$ and $B$ in figure 5.32 are either $+6$ V or $-5$ V, and $V_x$ is (a) $-2$ V, (b) $+3$ V, state the voltage which may be expected at the output terminal in each case. Neglect the effect of the forward p.d. in the diodes.

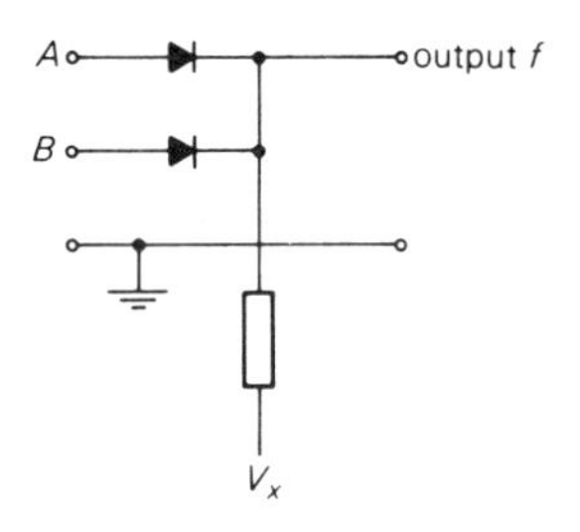


Figure 5.32

**5.3** Estimate suitable values for $R_1$ and $R_L$ in figure 5.6 given that $V_{CC} = +9$ V, $h_{FE(sat)} = 50$, $V_1 = 5$ V and that the maximum collector current is 10 mA. For the transistor, $V_{CE(sat)} = 0.15$ V and $V_{BE(sat)} = 0.6$ V.

**5.4** Explain what is meant by noise margin in connection with logic gates. Discuss means of increasing the noise immunity of logic circuits.

**5.5** Sketch the switching waveforms associated with an inverting logic gate and explain the reasons for the principal time delays.

**5.6** Explain the meaning of propagation delay in connection with a logic gate. Discuss methods of reducing the propagation delay in gates.

**5.7** Why does the use of an 'active' collector load improve the switching speed of an inverting gate? Give one advantage and one disadvantage of the use of diode D in the active load in figure 5.12.

**5.8** An AND gate having inputs $A$ and $B$ is connected by a wired-OR link to the output of a NOR gate with inputs $\overline{C}$ and $\overline{D}$. Deduce the logic function generated by the combination.

**5.9** What factors limit the number of inputs of a fan-in expander of the type in figure 5.15?

**5.10** Discuss the features of the branches of the TTL family and suggest applications for each branch of the family.

**5.11** Describe the operation of the CMOS NOR gate in figure 5.23a.

**5.12** Draw a cross-section through a monolithic integrated circuit which is equivalent to the circuit in figure 5.33. Show the interconnections between each of the circuit elements and the points A, B, C and D.

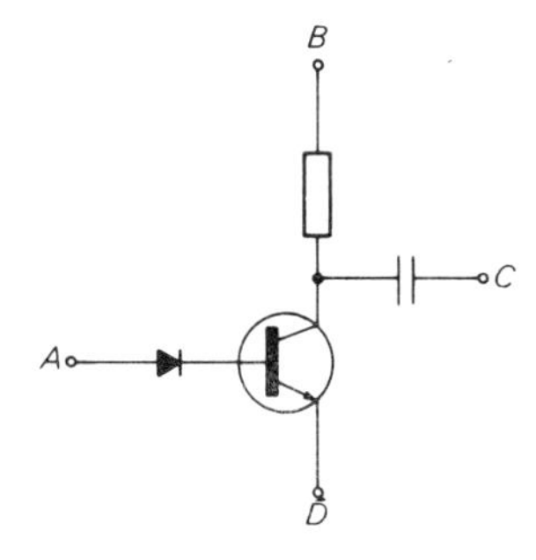


Figure 5.33

# 6 The Algebra of Logic

## 6.1 THE LAWS OF LOGIC

In order to test the truth or otherwise of logical statements we first draw up rules or laws of the processes involved. The principal laws are listed below.

### Commutative Law

This states that the order in which terms or variables appear in the equation is irrelevant.

$$A+B=B+A$$
$$A.B=B.A$$

### Associative Law

This states that the order in which identical functions are performed in the equation is irrelevant.

$$A+B+C=(A+B)+C=A+(B+C)$$
$$A.B.C=(A.B).C=A.(B.C)$$

*Note*: Care has been taken in bracketing terms together, since $A+B.(C+D)$ is not equal to $(A+B).(C+D)$, but it is equal to $A+(C+D).B$.

### Distributive Law

This is expressed in two forms

$$A+(B.C.D. \dots)=(A+B).(A+C).(A+D). \dots$$
$$A.(B+C+D+\dots)=A.B+A.C+A.D+\dots$$

The first gives the *product of sums* expression, and the second gives the *sum of products* result.

## 6.2 LOGIC THEOREMS

### De Morgan's Theorem

This states that the logical complement of a function is obtained if we (1) logically invert each term in the expression, and (2) interchange the 'dots' with the 'plusses' and vice versa, as follows.

$$\overline{A\,.\,B\,.\,C} = \overline{A} + \overline{B} + \overline{C}$$
$$\overline{A+B+C} = \overline{A}\,.\,\overline{B}\,.\,\overline{C}$$

### Other Useful Theorems

| | |
|---|---|
| (1) $A+0=A$ | (5) $A\,.\,0=0$ |
| (2) $A+1=1$ | (6) $A\,.\,1=A$ |
| (3) $A+A=A$ | (7) $A\,.\,A=A$ |
| (4) $A+\overline{A}=1$ | (8) $A\,.\,\overline{A}=0$ |
| | (9) $\overline{\overline{A}}=A$ |

## 6.3 APPLICATIONS OF THE LAWS AND THEOREMS OF LOGIC

A fundamental application of the laws and theorems outlined above is to the *analysis* and *simplification* of logical expressions. From such an analysis, it is possible to construct or *synthesise* logic networks; network synthesis is described in chapter 7. The following examples best illustrate the use of the laws and theorems.

*Example 6.1*

Signals from sensors *A*, *B* and *C* in a security system must give an indication of the presence of an intruder when the following logical conditions are satisfied

$$f = A\,.\,B\,.\,C + A\,.\,\overline{B}\,.\,C + A\,.\,B\,.\,\overline{C}$$

Simplify the expression for *f*.

*Solution* The equation tells us that an alarm signal is initiated when any or all of the three logical groupings on the right-hand side of the equation are satisfied.

Inspecting the terms in the equation, we see that the term $A\,.\,B$ appears in the first and third groups. First then, let us apply the commutative law to collect like terms together, and then apply the distributive law as follows

$$f = A\,.\,B\,.\,C + A\,.\,B\,.\,\overline{C} + A\,.\,\overline{B}\,.\,C \qquad \text{(commutative law)}$$
$$= A\,.\,B\,.\,(C+\overline{C}) + A\,.\,\overline{B}\,.\,C \qquad \text{(distributive law)}$$

We now apply theorem 4 to simplify the first term in the expression

$$f = A\,.\,B\,.\,1 + A\,.\,\overline{B}\,.\,C \qquad \text{(theorem 4)}$$

Applying theorem 6 we find that

$$f = A\,.\,B + A\,.\,\overline{B}\,.\,C$$

From the second form of the distributive law, we can further say that

$$f = A\,.\,(B + \overline{B}\,.\,C) \qquad (6.1)$$

This is one possible form of solution. Another form can be deduced from the fact that the term $A\,.\,C$ appears in the first and second expressions of the original equation. The equation is then simplified as follows.

$$f = A\,.\,C\,.\,B + A\,.\,C\,.\,\overline{B} + A\,.\,B\,.\,\overline{C} \qquad \text{(commutative law applied twice)}$$
$$= A\,.\,C\,.\,(B+\overline{B}) + A\,.\,B\,.\,\overline{C} \qquad \text{(distributive law)}$$
$$= A\,.\,C\,.\,1 + A\,.\,B\,.\,\overline{C} \qquad \text{(theorem 4)}$$
$$= A\,.\,C + A\,.\,B\,.\,\overline{C} \qquad \text{(theorem 6)}$$
$$= A\,.\,(C + B\,.\,\overline{C}) \qquad (6.2)$$

Although equations 6.1 and 6.2 differ slightly, they both represent the initial equation.

In the above solutions we have used different sections of the term $A\,.\,B\,.\,C$ in the simplification process. This leads to the idea of another form of solution, since by using theorem 3 in reverse we can say that

$$A\,.\,B\,.\,C = A\,.\,B\,.\,C + A\,.\,B\,.\,C$$

If this is substituted into the original expression we get

$$f = A\,.\,B\,.\,C + A\,.\,B\,.\,C + A\,.\,\overline{B}\,.\,C + A\,.\,B\,.\,\overline{C} \quad \text{(theorem 3)}$$

$$= (A\,.\,B\,.\,C + A\,.\,B\,.\,\overline{C}) + (A\,.\,B\,.\,C + A\,.\,\overline{B}\,.\,C) \quad \text{(commutative law)}$$

$$= A\,.\,B\,.\,(C + \overline{C}) + A\,.\,C\,.\,(B + \overline{B}) \quad \text{(distributive law)}$$
$$= A\,.\,B\,.\,1 + A\,.\,C\,.\,1 \quad \text{(theorem 4)}$$
$$= A\,.\,B + A\,.\,C \quad \text{(theorem 6)}$$
$$= A\,.\,(B + C) \quad \text{(distributive law)}$$

Once more we have arrived at a solution which differs from equations 6.1 and 6.2. The solution just obtained is, in fact, the *minimal solution* expressed in terms of a logical algebraic expression. However, the expression 'minimal solution' can be interpreted in several ways and the *logical* minimal expression is not necessarily the best solution from a circuit viewpoint. For example, we have obtained a number of possible versions of the original equation, and while the third version gives the neatest logical solution, it remains to be seen whether it provides the best electrical solution in terms of the number of IC packs required, or of the number and length of interconnections involved, or of the overall speed of operation, or of the overall cost. The solution adopted is often a compromise between these factors.

*Example 6.2*

Tests on an integrated circuit reveal that it satisfies the truth table given below. Determine the logical function the IC generates, and simplify the expression as far as possible.

| *Inputs* $A$ | $B$ | $C$ | *Output* $X$ |
|---|---|---|---|
| 0 | 0 | 0 | 0 |
| 0 | 0 | 1 | 0 |
| 0 | 1 | 0 | 1 |
| 0 | 1 | 1 | 0 |
| 1 | 0 | 0 | 0 |
| 1 | 0 | 1 | 1 |
| 1 | 1 | 0 | 1 |
| 1 | 1 | 1 | 1 |

*Solution* One method of deriving the function generated is to write down the logical equation which gives *all* the conditions which provide a '1' at the output. The first condition occurs in the third row, when

$$A = 0\,(\text{that is, } \overline{A} = 1)\ \text{AND}\ B = 1\ \text{AND}\ C = 0\,(\text{that is, } \overline{C} = 1).$$

Thus $X = 1$ when the input condition $\overline{A}\,.\,B\,.\,\overline{C} = 1$ is satisfied. Progressing down the list we see that $X$ is also '1' when the conditions $A\,.\,\overline{B}\,.\,C$, $A\,.\,B\,.\,\overline{C}$, $A\,.\,B\,.\,C$ are satisfied. This is written down in logical form as

$$X = \overline{A}\,.\,B\,.\,\overline{C} + A\,.\,\overline{B}\,.\,C + A\,.\,B\,.\,\overline{C} + A\,.\,B\,.\,C$$

This expression, although complete, has yet to be minimised. To minimise the expressions we group like terms together as far as possible. This we do by grouping the first and third terms, and also the second and fourth terms.

$$X = (\overline{A}\,.\,B\,.\,\overline{C} + A\,.\,B\,.\,\overline{C}) + (A\,.\,\overline{B}\,.\,C + A\,.\,B\,.\,C)$$
$$= B\,.\,\overline{C}\,.\,(\overline{A} + A) + A\,.\,C\,.\,(\overline{B} + B)$$
$$= B\,.\,\overline{C} + A\,.\,C \quad (6.3)$$

Equation 6.3 is one form of minimal expression which satisfies the truth table.

An alternative approach is to say that *all* the conditions defined by 1s in the truth table are simply *NOT the* 0s. As the 1s in the truth table define $X$, then the zeros in the truth table define the function $\overline{X}$, so that

$$\overline{X} = \overline{A}\,.\,\overline{B}\,.\,\overline{C} + \overline{A}\,.\,\overline{B}\,.\,C + \overline{A}\,.\,B\,.\,C + A\,.\,\overline{B}\,.\,\overline{C}$$

By grouping the first term in the expression for $\overline{X}$ with the fourth term and the second term with the third term, we see that

$$\begin{aligned}\overline{X} &= (\overline{A}\,.\,\overline{B}\,.\,\overline{C} + A\,.\,\overline{B}\,.\,\overline{C}) + (\overline{A}\,.\,\overline{B}\,.\,C + \overline{A}\,.\,B\,.\,C)\\ &= \overline{B}\,.\,\overline{C}\,.\,(\overline{A} + A) + \overline{A}\,.\,C\,.\,(\overline{B} + B)\\ &= \overline{B}\,.\,\overline{C} + \overline{A}\,.\,C \qquad (6.4)\end{aligned}$$

Now, from theorem 9, $\overline{\overline{X}} = X$ and, if we complement (that is, NOT) both sides of equation 6.4 we arrive at an equation for $X$ of

$$X = \overline{\overline{X}} = \overline{\overline{B}\,.\,\overline{C} + \overline{A}\,.\,C} \qquad (6.5)$$

Equation 6.5 is another minimal form of the original equation. Although equations 6.3 and 6.5 differ in appearance they are, in fact, equivalent to one another as we shall show in the following.

In equation 6.5 let $M = \overline{B}\,.\,\overline{C}$ and $N = \overline{A}\,.\,C$. Applying De Morgan's theorem to equation 6.5 we see that

$$X = \overline{M + N} = \overline{M}\,.\,\overline{N} = (\overline{\overline{B}\,.\,\overline{C}})\,.\,(\overline{\overline{A}\,.\,C})$$

Applying De Morgan's theorem once more gives

$$\begin{aligned}X &= (B + C)\,.\,(A + \overline{C}) && \text{(De Morgan)}\\ &= A\,.\,B + B\,.\,\overline{C} + A\,.\,C + C\,.\,\overline{C} && \text{(distributive law)}\\ &= A\,.\,B + B\,.\,\overline{C} + A\,.\,C + 0 && \text{(theorem 8)}\\ &= A\,.\,B + B\,.\,\overline{C} + A\,.\,C && \text{(theorem 1)}\\ &= A\,.\,B\,.\,(C + \overline{C}) + (B\,.\,\overline{C} + B\,.\,\overline{C}) + (A\,.\,C + A\,.\,C) && \text{(theorems 3 and 4)}\\ &= (A\,.\,B\,.\,C + A\,.\,B\,.\,\overline{C} + B\,.\,\overline{C} + A\,.\,C) + (B\,.\,\overline{C} + A\,.\,C)\\ &= (A\,.\,C\,.\,(B + 1) + B\,.\,\overline{C}\,.\,(A + 1)) + (B\,.\,\overline{C} + A\,.\,C)\\ &= (A\,.\,C + B\,.\,\overline{C}) + (B\,.\,\overline{C} + A\,.\,C) && \text{(theorems 2 and 6)}\\ &= A\,.\,C + B\,.\,\overline{C} && \text{(theorem 3)}\end{aligned}$$

which is equivalent to equation 6.3

*Example 6.3*

In designing a logic network which controls part of a computer switching system, the following function must be satisfied

$$f = W\,.\,(\overline{\overline{X} + \overline{W}\,.\,(Y + \overline{X}\,.\,\overline{Y})}) \qquad (6.6)$$

Simplify the expression.

*Solution* The first step is to write the equation in the form $f = W\,.\,D$, where

$$D = \overline{\overline{X} + \overline{W}\,.\,(Y + \overline{X}\,.\,\overline{Y})}.$$

Expanding the expression under the 'bar' we get

$$D = \overline{\overline{X} + \overline{W}\,.\,Y + \overline{W}\,.\,\overline{X}\,.\,\overline{Y}}$$

Further, if we let $E = \overline{W}\,.\,Y$, and $F = \overline{W}\,.\,\overline{X}\,.\,\overline{Y}$, then

$$\begin{aligned}D &= \overline{\overline{X} + E + F} = X\,.\,\overline{E}\,.\,\overline{F} && \text{(De Morgan)}\\ &= X\,.\,(\overline{\overline{W}\,.\,Y})\,.\,(\overline{\overline{W}\,.\,\overline{X}\,.\,\overline{Y}})\end{aligned}$$

Now, by De Morgan's theorem

$$\overline{\overline{W}\,.\,Y} = W + \overline{Y}$$

and

$$\overline{\overline{W}\,.\,\overline{X}\,.\,\overline{Y}} = W + X + Y$$

Hence

$$\begin{aligned}D &= X.(W+\bar{Y}).(W+X+Y)\\ &= (W.X+X.\bar{Y}).(W+X+Y)\\ &= W.X+W.X+W.X.Y+W.X.\bar{Y}+X.\bar{Y}\\ &\quad +X.Y.\bar{Y}\\ &= W.X+X.\bar{Y}\end{aligned}$$

Now

$$\begin{aligned}f = W.D &= W.(W.X+X.\bar{Y})\\ &= W.(W.X.(Y+\bar{Y})+X.\bar{Y}.(W+\bar{W}))\\ &= W.(W.X.Y+W.X.\bar{Y}+\bar{W}.X.\bar{Y})\\ &= W.X.Y+W.X.\bar{Y}+W.\bar{W}.X.\bar{Y}\\ &= W.X.(Y+\bar{Y}) = W.X\end{aligned}$$

That is, the complex logical function described by equation 6.6 is simply equivalent to $W$ AND $X$, that is, input $Y$ is *redundant* and plays no part in the operation of the system.

## PROBLEMS

**6.1** Use truth tables to verify that (a) $A+B.C.D=(A+B).(A+C).(A+D)$, (b) $A.(B+C+D)=A.B+A.C+A.D$.

**6.2** Use truth tables to verify De Morgan's theorem (a) $\overline{A.B}=\bar{A}+\bar{B}$, (b) $\overline{A+B}=\bar{A}.\bar{B}$.

**6.3** Show that $A.\bar{B}.\bar{C}+B.C+B.\bar{C}=B+A.\bar{C}$.

**6.4** Prove by means of Boolean algebra that $A.\bar{C}+B+C.\bar{B}=A+B+C$.

**6.5** Where possible, simplify the following logical expressions (a) $B+A.B$, (b) $A+1$, (c) $A.0$, (d) $B.(A+B)$, (e) $A.\bar{B}+\bar{A}.B$.

**6.6** Show that (a) $\overline{A.B.\bar{C}}=\bar{A}+\bar{B}+C$, (b) $\overline{C.(A+\bar{B})}=\bar{C}+\bar{A}.B$, (c) $A.B.\overline{(\overline{C.(A+\bar{B})}+A.B.\bar{C})}=A.B.C$.

**6.7** Prove that $\overline{\overline{(\bar{B}.C+D)}.(\bar{A}+\bar{B})}.(\bar{A}.D)=\bar{A}.D$.

**6.8** Which of the following is $\bar{A}.(A+B)$ equal to: (a) $\bar{A}$, (b) $A.B$, (c) $\bar{A}.B$, (d) $A+B$?

**6.9** Which is the redundant term in the expression $A.\bar{C}+A.B+B.C+\bar{A}.C$?

**6.10** Which of the following is equal to logic '0': (a) $A+1$, (b) $A.\bar{A}$, (c) $A+A.\bar{A}$, (d) $\bar{A}+1$, (e) $(\bar{A}+\bar{B}).B.(A+\bar{A}.\bar{B})$?

**6.11** Show that $A.\bar{B}.C.\bar{D}+A.B.C+\bar{A}.C.D+A.C.D+C.D=C.(D+A)$.

# 7 Logic System Design

## 7.1 COMBINATIONAL LOGIC AND SEQUENTIAL LOGIC NETWORKS

Logic networks fall into two broad classes, namely *combinational logic networks* and *sequential logic networks.* Combinational logic networks include the networks dealt with so far, and which generate an output signal when a given combination or combinations of input signals exist. The output from sequential logic systems depend on the sequence of events which have already occurred in the circuit, and include such systems as counters. In this chapter we deal with the design of combinational logic systems.

## 7.2 THE DESIGN OF LOGIC NETWORKS FROM TRUTH TABLES

As we have seen in example 6.2, we can deduce the logical equation which completely defines a truth table, and in the following we shall see how to design a logic system from the equation. Consider the equation in table 7.1.

**Table 7.1**

| *Inputs* *A* | *B* | *C* | *Output* *X* |
|---|---|---|---|
| 0 | 0 | 0 | 0 |
| 0 | 0 | 1 | 1 |
| 0 | 1 | 0 | 0 |
| 0 | 1 | 1 | 0 |
| 1 | 0 | 0 | 1 |
| 1 | 0 | 1 | 1 |
| 1 | 1 | 0 | 0 |
| 1 | 1 | 1 | 0 |

Writing down the expression which defines the 1s in the table gives

$$X = \overline{A}\,.\,\overline{B}\,.\,C + A\,.\,\overline{B}\,.\,\overline{C} + A\,.\,\overline{B}\,.\,C \tag{7.1}$$

If the sensors which energise the logic system detect the state of the variables $A$, $B$ and $C$, then we need three NOT gates to generate the functions $\overline{A}$, $\overline{B}$ and $\overline{C}$ required in equation 7.1. These gates are shown in figure 7.1a.

The function $\overline{A} . \overline{B} . C$ is generated by a 3-input AND gate, its inputs being energised by signals $\overline{A}$, $\overline{B}$ and $C$ in the manner shown in figure 7.1b. The output from this gate is designated the letter $L$. In equation 7.1 the terms $A . \overline{B} . \overline{C}$ $(= M)$ and $A . \overline{B} . C$ $(= N)$ are generated as shown in figure 7.1c.

Finally the function $X$ is generated by OR-gating the terms $L$, $M$, and $N$ as shown in figure 7.1d. The completed block diagram is shown in figure 7.1e, in which the individual sections are linked together.

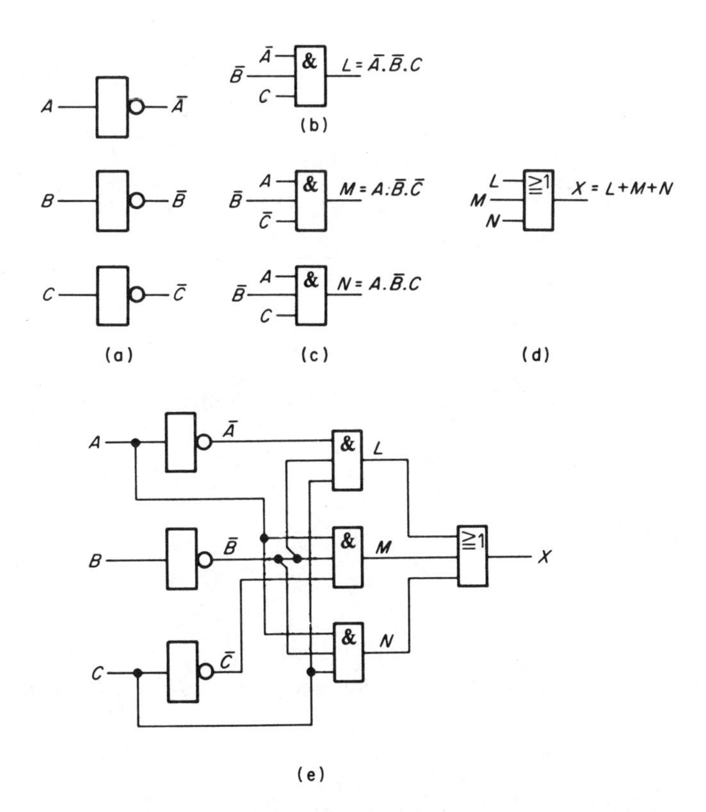


Figure 7.1 The elements generating the basic functions in equation 7.1

Using the techniques outlined in chapter 6, equation 7.1 can be simplified to

$$X = A . \overline{B} + C . \overline{B} \tag{7.2}$$
$$= \overline{B} . (A + C) \tag{7.3}$$

The block diagram of the logic network which satisfies equation 7.2 is shown in figure 7.2, and requires four gates. The logic block diagram corresponding to equation 7.3 is shown in figure 7.3 and only three gates are required in this case.

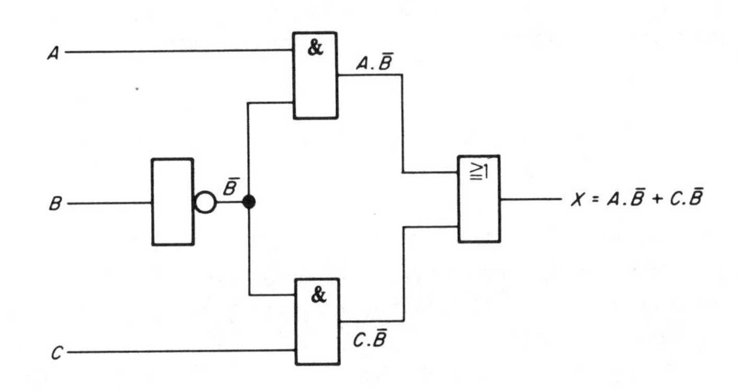


Figure 7.2 A logic block diagram satisfying equation 7.2

An alternative approach is to develop an equation which defines NOT the 0s in the truth table. Such an equation is

$$\overline{X} = \overline{A} . \overline{B} . \overline{C} + \overline{A} . B . \overline{C} + \overline{A} . B . C + A . B . \overline{C} + A . B . C$$

Applying De Morgan's theorem to the above equation gives

$$X = (A + B + C) . (A + \overline{B} + C) . (A + \overline{B} + \overline{C}) . (\overline{A} + \overline{B} + C) . (\overline{A} + \overline{B} + \overline{C})$$

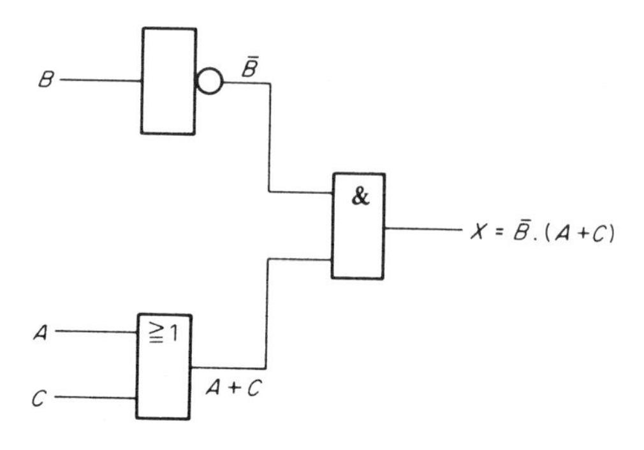


Figure 7.3 A logic block diagram satisfying equation 7.3

Although this is a complex logical statement, it is well suited for direct implementation in the form of a NOR network.

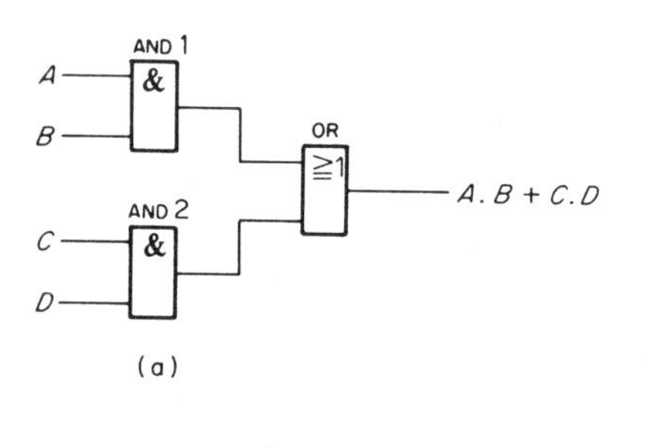


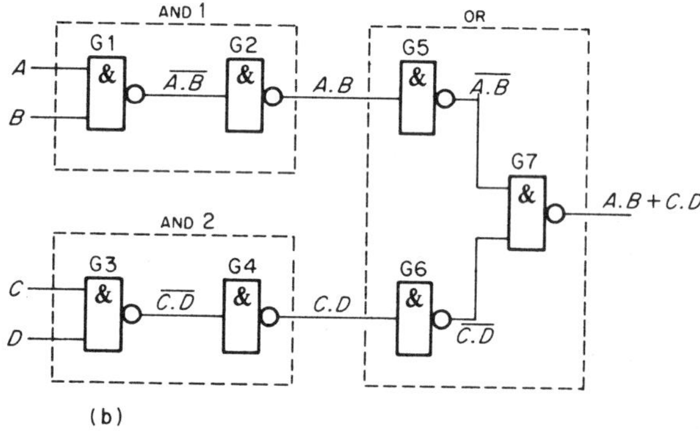


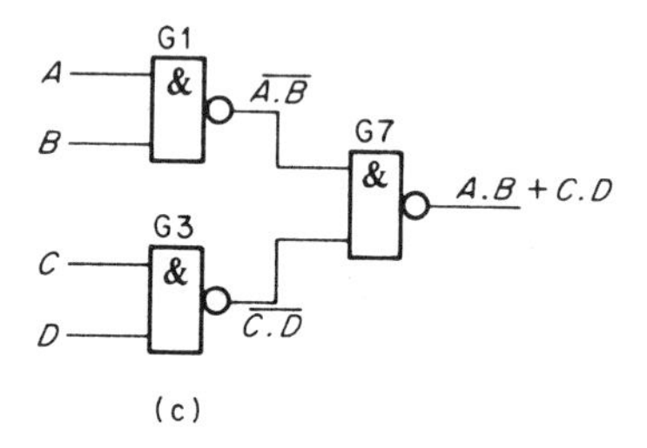


Figure 7.4 Generating the function $f = A\,.\,B + C\,.\,D$ using NAND gates

## 7.3 REPLACING AND–OR NETWORKS BY NAND NETWORKS

When we write down the logical equation defining all the 1s in a truth table, we obtain an equation which is in the *sum of products* form, that is, in an AND–OR type of equation. An example of this type was given in equation 7.1 of section 7.2. To illustrate the basic principles, we shall consider the design of a NAND network which generates the function $A\,.\,B + C\,.\,D$. Using AND and OR gates, the network required is shown in figure 7.4a. To replace figure 7.4a by NAND elements, we first replace each AND element by its NAND equivalent (see figure 3.3b), and replace the OR element by the NAND equivalent as in figure 3.3b. The resulting NAND network is shown in figure 7.4b. Now, using the minimisation technique in figure 3.4a, we remove gates G2 and G5 in figure 7.4b and replace them by a single wire, and similarly we also replace gates G4 and G6. This leaves only gates G1, G2 and G7 in figure 7.4b, and the simplified NAND network is shown in figure 7.4c.

Comparing circuits a and c in figure 7.4 we see that the AND–OR network is replaced by an equivalent all-NAND network. This technique is applicable to any AND–OR type of network, and you may like to show that if we replace each gate in figure 7.1e by a NAND gate then the output of the resulting all-NAND network is given by equation 7.1.

## 7.4 REPLACING OR–AND NETWORKS WITH NOR NETWORKS

It was shown in section 7.2 that the logical equation which defines all the 0s in a truth table can be written in the *logical product-of-sums* form, that is, in an OR–AND form such as $(A + B) . (C + D)$.

Let us consider a network which generates the function $(A + B) . (C + D)$. The basic OR–AND network is shown in figure 7.5a and, using the circuits in figure 3.5, it is implemented in NOR form in figure 7.5b. Note that gates G2 and G5 form a cascaded pair of NOR gates which have a single input and are therefore redundant, as are gates G4 and G6. This leaves gates G1, G3 and G7 to generate the required function. The resulting NOR network is shown in figure 7.5c.

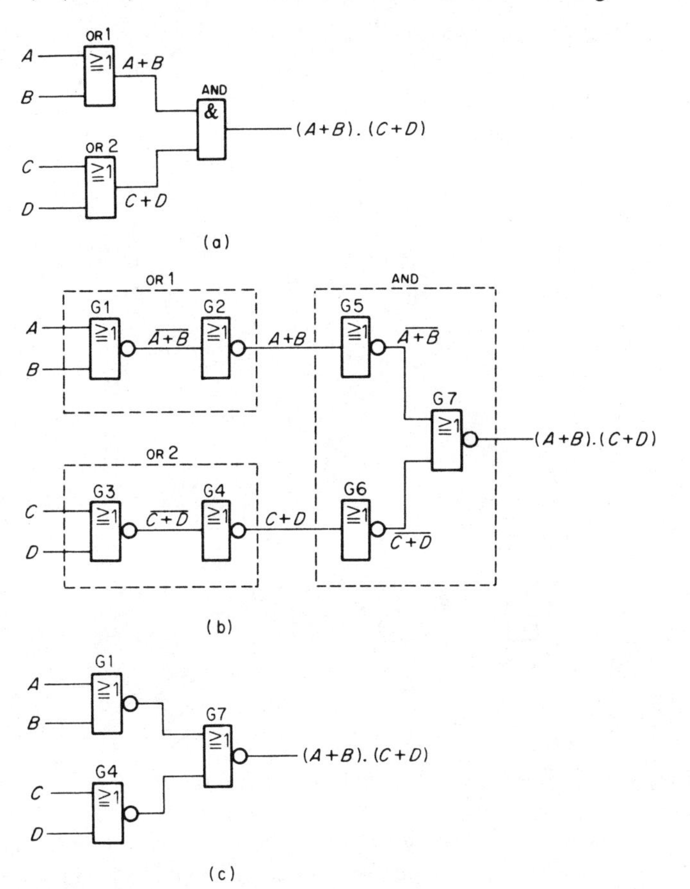


Figure 7.5 Generating the function $f = (A + B) . (C + D)$ using NOR gates

## 7.5 WIRED-OR NETWORKS

The wired-OR connection (see also section 5.12) can be used with either NAND or NOR gates provided that they have pull-up resistors in their output circuits. It was shown in section 5.12 that the function generated by this connection could be regarded as the AND function of the individual outputs from the gates. To illustrate the effect on NOR and NAND networks, we shall consider each separately.

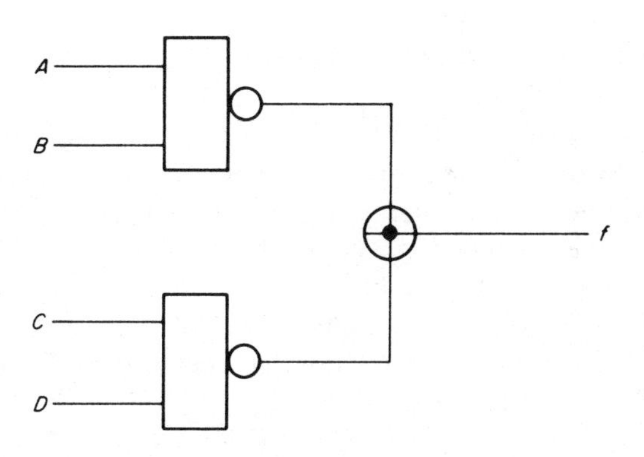


Figure 7.6 The wired-OR connection

### NOR Networks

Two gates connected by a wired-OR link are shown in figure 7.6. The logical expression for output $f$ is

$$f = \overline{(A + B)} . \overline{(C + D)}$$

$$= (\overline{A} \,.\, \overline{B}) \,.\, (\overline{C} \,.\, \overline{D})$$
$$= \overline{A} \,.\, \overline{B} \,.\, \overline{C} \,.\, \overline{D}$$
$$= \overline{A + B + C + D}$$

That is, when NOR gates are used in wired-OR networks the resulting function generated is the NOR function of all the input signals. Hence the wired-OR connection can be used as a means of increasing the fan-in of NOR networks.

### NAND Networks

If we use the network in figure 7.6 once more, but this time using NAND gates, the function generated is

$$f = \overline{(A \,.\, B)} \,.\, \overline{(C \,.\, D)}$$
$$= \overline{A \,.\, B + C \,.\, D}$$

## 7.6 THE EXCLUSIVE-OR FUNCTION

Gates generating the EXCLUSIVE-OR function are widely used in logic circuits for operations including addition, subtraction, multiplication, division, and binary number comparison. The truth table defining the function is given in table 7.2.

The first three rows of table 7.2 are identical to those of the 2-variable OR function (see also table 2.2). Table 7.2 differs from that of the OR function only in the final line since, in table 7.2, when $A = B = 1$ the output is logic '0'.

**Table 7.2**

| *Inputs* | | *Output* |
|---|---|---|
| $A$ | $B$ | $S$ |
| 0 | 0 | 0 |
| 0 | 1 | 1 |
| 1 | 0 | 1 |
| 1 | 1 | 0 |

In this part of the book we shall consider the design of networks which generate the EXCLUSIVE-OR function, applications of the circuit being introduced at appropriate points throughout the book. Inspecting table 7.2, we see that output $S$ has the value '1' when $A \,.\, \overline{B} = 1$ or when $\overline{A} \,.\, B = 1$, that is

$$S = A \,.\, \overline{B} + \overline{A} \,.\, B \tag{7.4}$$

This function is implemented in figure 7.7a using AND, OR and NOT gates. The term $A \,.\, \overline{B}$ is generated by G1, $\overline{A} \,.\, B$ is generated by G2, and G3 forms the final output. Two symbols used to represent EXCLUSIVE-OR gates are shown in figure 7.7b.

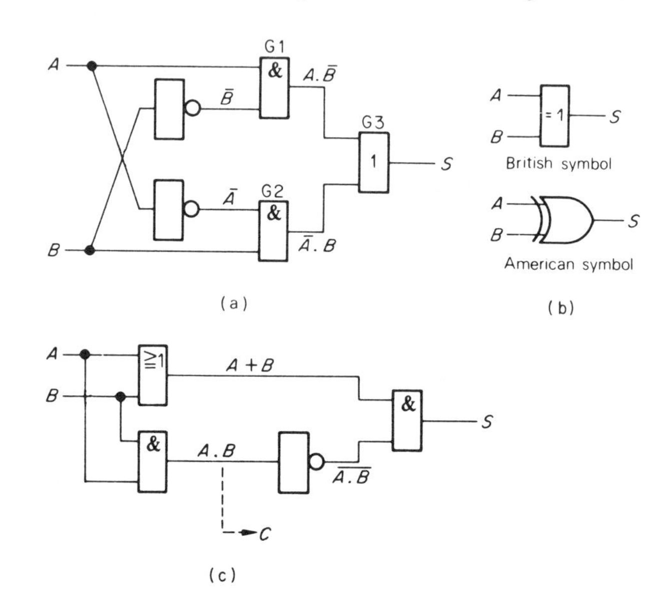


Figure 7.7 The EXCLUSIVE-OR function

An EXCLUSIVE-OR circuit which requires only four gates can be developed by applying our knowledge of Boolean algebra to the equation for $S$ as follows.

$$S = A \,.\, \overline{B} + \overline{A} \,.\, B + 0 + 0 \qquad \text{(theorem 1)}$$

$$= A \,.\, \overline{B} + \overline{A} \,.\, B + A \,.\, \overline{A} + B \,.\, \overline{B} \qquad \text{(theorem 8)}$$
$$= (A + B) \,.\, (\overline{A} + \overline{B}) \qquad (7.5)$$
$$= (A + B) \,.\, (\overline{A \,.\, B}) \qquad (7.6)$$

The block diagram of a circuit which generates equation 7.6 is shown in figure 7.7c. The circuit also generates at point C the function $A \,.\, B$. This additional output is particularly useful in arithmetic circuits and is discussed further in chapter 9.

Equation 7.5 can also be written in the form

$$S = A \,.\, (\overline{A} + \overline{B}) + B \,.\, (\overline{A} + \overline{B})$$
$$= A \,.\, (\overline{A \,.\, B}) + B \,.\, (\overline{A \,.\, B})$$

This function is generated by the NAND network in figure 7.8, since the output from this circuit is

$$S = \overline{\overline{(A \,.\, (\overline{A \,.\, B}))} \,.\, \overline{(B \,.\, (\overline{A \,.\, B}))}}$$
$$= A \,.\, (\overline{A \,.\, B}) + B \,.\, (\overline{A \,.\, B})$$

The circuit in figure 7.8 can be constructed using a single quad 2-input NAND IC package (see figure 5.31).

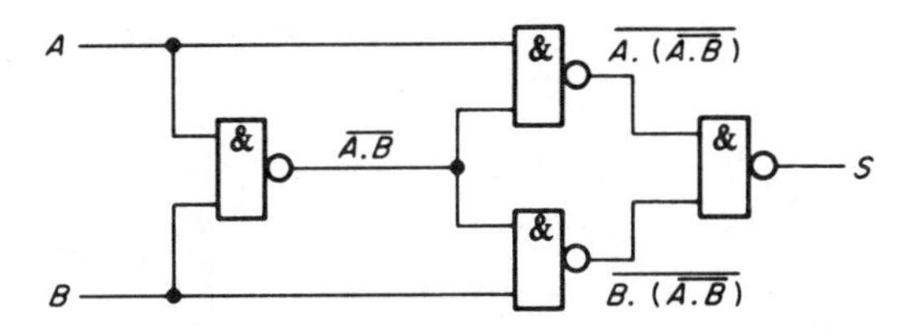


Figure 7.8 Generating the EXCLUSIVE-OR function using four 2-input NAND gates

The EXCLUSIVE-OR gate is also known as the *non-equivalence* gate since from table 7.2, the output is logic '1' when the inputs are logically not equivalent to one another, that is, when $A = 1$, $B = 0$ or vice versa. It is also known as a *modulo-2* addition circuit since the output is '1' when one input only is energised, otherwise the output is zero.

## 7.7 KARNAUGH MAPS

So far we have attempted to minimise networks using Boolean algebra only. In some instances the reason for steps taken during the minimisation procedure is not always obvious, and it often takes a little time to obtain a satisfactory solution. An alternative minimisation technique uses what is known as a Karnaugh map, named after the scientist who devised it. The Karnaugh map is simply a method of plotting or mapping all the conditions given in the truth table.

### Single Variable Map

A single variable can have one of two possible operating states; if the variable is designated the symbol $A$, then it has either the logical value '1' ($A = 1$), or it has the value '0' ($\overline{A} = 1$). In the case of a single variable, the Karnaugh map contains two equal divisions, known as *cells*, as shown in figure 7.9a, one half of the map representing the state of $A$ and the other half representing $\overline{A}$. If we wish to represent the condition $A = 1$, we do so by drawing the map in figure 7.9b. The $\overline{A}$ cell contains zero, since, when $A = 1$, variable $\overline{A} = 0$. We represent the condition $\overline{A} = 1$ (that is, $A = 0$) by the map in figure 7.9c.

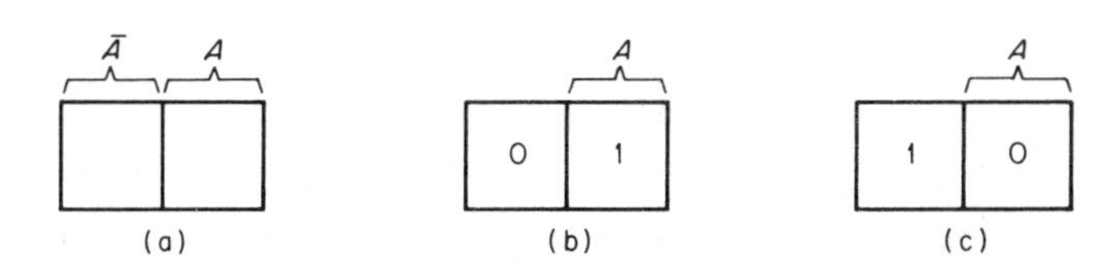


Figure 7.9 Karnaugh map for a single variable

### A 2-variable Map

Each variable associated with a problem has two possible states, so that when there are two variables we have four possible combinations of the variables. Hence the 2-variable maps in figure 7.10 have four cells to represent these states. There are two popular methods of drawing Karnaugh maps, both being shown in figure 7.10 and each having its advantages. Let us consider the map in figure 7.10a.

As before, each variable must be represented by one-half of the total number of cells, and in this case we allow variable $A$ to represent the cells in the right-hand half of the map. The two cells in the left-hand column represent $\bar{A}$. The two lower cells represent the variable $B$ and the two upper cells represent $\bar{B}$. Each cell in the map is defined by the *intersection* or *union* of the variables, much as a point on a geographical map can be defined by horizontal and vertical references. Thus the cell in the lower right-hand corner is the intersection of the variables $A$ and $B$, and is shown as the cell $A . B$. The cell directly above it is located by the intersection of variables $A$ and (NOT $B$), and is shown as the cell $A . \bar{B}$. The other cells on the map are defined in much the same way.

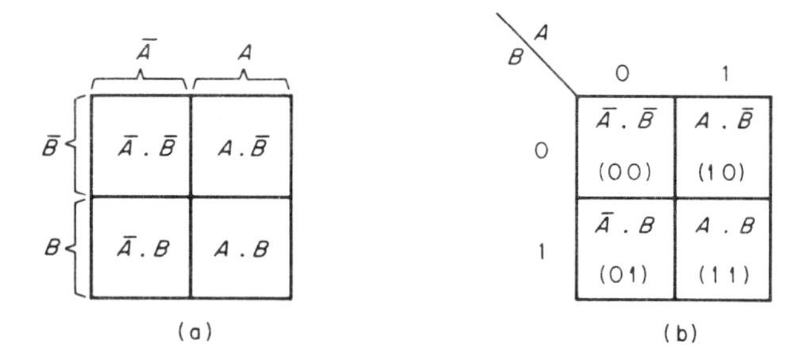


Figure 7.10 Karnaugh map for two variables

In the map in figure 7.10b, we represent variable $A$ by placing a '1' at the head of the column representing $A$ and a '0' at the head of the $\bar{A}$ column. Similarly a '1' placed at the left-hand end of the lower row of cells shows that that row represents variable $B$, and a '0' at the left-hand end of the upper row indicates that it represents $\bar{B}$. The binary grouping written inside the cells (that is, 00, 10, 01, 11) represents the values of the *input* variables associated with the cells. For example, when $A = 0$ and $B = 1$ the cell concerned is defined by the binary group 01, that is, the cell $\bar{A} . B$. We shall now illustrate how the Karnaugh map is used in problems involving logic.

Let us consider how we map the function given in table 7.3. The Karnaugh map associated with table 7.3 is illustrated in figure 7.11a, and is formed as follows. Considering the table one row at a time, the input conditions corresponding to the first row are $A = 0$, $B = 0$, that is, we are concerned with the cell $\bar{A} . \bar{B}$ on the map. We place in this cell the value of the function $f$ which is the output from the network. Since $f = 0$ in this case we place a '0' in the upper left-hand cell (cell $\bar{A} . \bar{B}$) of the map in figure 7.11a. Similarly, cells $\bar{A} . B$ and $A . \bar{B}$ have 0s placed in them. Since a '1' appears in the final row of the $f$ column of the truth table, a '1' must be written in the lower right-hand cell (cell $A . B$) of the Karnaugh map.

The map in figure 7.11b is formed in much the same way. Cell 00 (the upper left-hand cell) is defined by the input conditions $A = 0$, $B = 0$ and in the truth table we note that $f = 0$ for this state.

**Table 7.3**

| *Inputs* | | *Output* | *Cell defined* |
|---|---|---|---|
| *A* | *B* | *f* | *by the inputs* |
| 0 | 0 | 0 | $\bar{A} . \bar{B}$ |
| 0 | 1 | 0 | $\bar{A} . B$ |
| 1 | 0 | 0 | $A . \bar{B}$ |
| 1 | 1 | 1 | $A . B$ |

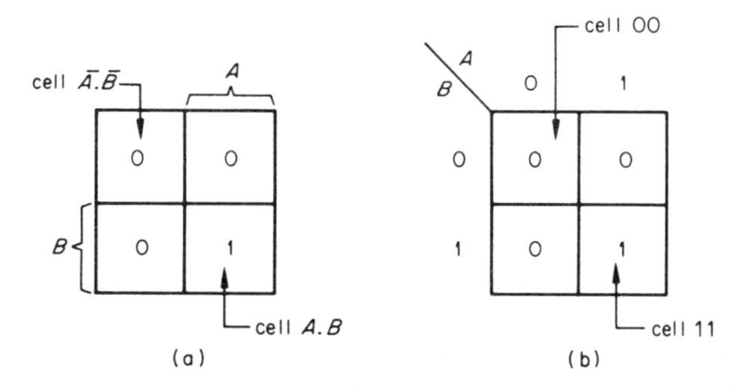


Figure 7.11 Karnaugh map for the function $f = A . B$

Accordingly we write a '0' in that cell. Cell 01 (the lower left-hand cell) corresponds to the input conditions $A = 0$, $B = 1$, for which $f = 0$; a '0' is written in that cell. In cell 11 (the lower right-hand cell) we record a '1' since $f = 1$ when $A = 1$ and $B = 1$.

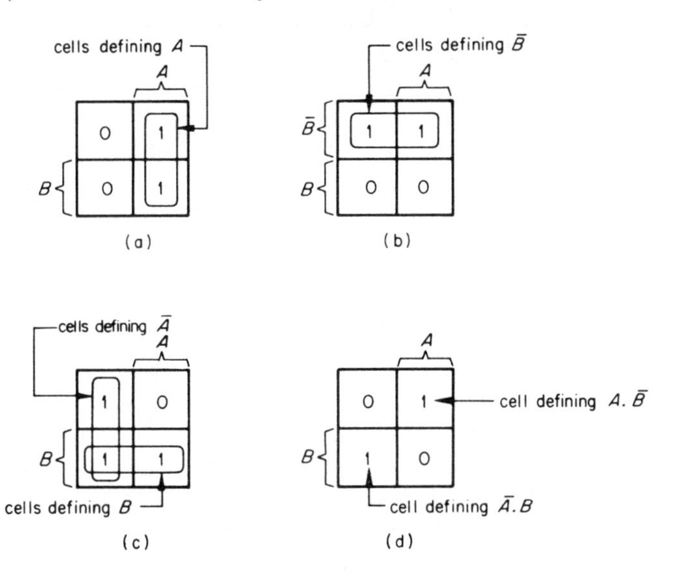


Figure 7.12 Grouping adjacent cells on a two-variable map

Now suppose we have 1s in more than one cell, as shown in figure 7.12a, in which the cells defined by 1s are $A\,.\,B$ and $A\,.\,\overline{B}$. A logical network which generates an output corresponding to the map in figure 7.12a provides a '1' output either when $A\,.\,B = 1$ OR when $A\,.\,\overline{B} = 1$, that is

$$f = A\,.\,B + A\,.\,\overline{B}$$

Applying the rules of Boolean algebra to the above equation, we simplify it as follows

$$f = A\,.\,(B + \overline{B}) = A\,.\,1 = A$$

That is, the map in figure 7.12a defines the function $f = A$. We can derive the same result from the Karnaugh map by grouping together *adjacent cells* which contain 1s in the manner shown in figure 7.12a. *Adjacent cells are defined as cells which differ in their binary representation by only one binary digit (bit).* For example, cell $A\,.\,B$ is represented by the binary input group 11, and cell $A\,.\,\overline{B}$ by the group 10; these differ only in the value of the right-hand bit. More will be said about this later.

The map in figure 7.12b is that of the logical equation

$$f = \overline{A}\,.\,\overline{B} + A\,.\,\overline{B}$$

which, by Boolean algebra, can be reduced to

$$f = \overline{B}\,.\,(\overline{A} + A) = \overline{B}$$

Once more, we can group two adjacent cells on the Karnaugh map in order to simplify the defining equation, as shown in figure 7.12b.

The map in figure 7.12c includes 1s in three cells, and the logical equation satisfied by the map is

$$f = A\,.\,B + \overline{A}\,.\,B + \overline{A}\,.\,\overline{B} \tag{7.7}$$

This equation can be simplified to $f = \overline{A} + B$ by Boolean algebra, but the steps involved are not obvious at first sight. However, if we group the cells containing 1s in the manner shown in figure 7.12c, we see that the horizontal group of cells includes *all* cells defined by variable $B$ and the vertical group of cells includes *all* cells defined by $\overline{A}$, hence

$$f = \overline{A} + B$$

This introduces us to the concept that any '1' on the Karnaugh map can be used several times, a fact which is proved in the following. From theorem 3 (chapter 6), we see that

$$\overline{A}\,.\,B + \overline{A}\,.\,B = \overline{A}\,.\,B$$

that is, the term $\overline{A}\,.\,B$ (or any other term for that matter) may be repeated as *many times* as we like (the process is not limited simply to its use twice) once it has appeared in our equation. Thus, if we repeat $\overline{A}\,.\,B$ twice in equation 7.7, we get

$$
\begin{aligned}
f &= A\,.\,B+(\overline{A}\,.\,B+\overline{A}\,.\,B)+\overline{A}\,.\,\overline{B}\\
&= (A\,.\,B+\overline{A}\,.\,B)+(\overline{A}\,.\,B+\overline{A}\,.\,\overline{B}) \qquad (7.8)\\
&= B\,.\,(A+\overline{A})+\overline{A}\,.\,(B+\overline{B}) = B+\overline{A}
\end{aligned}
$$

The first bracketed pair of terms in equation 7.8 corresponds to the lower pair of cells in figure 7.12c, while the second bracketed pair of terms corresponds to the pair of cells in the left-hand column of figure 7.12c.

An example of a function which cannot be simplified is mapped in figure 7.12d. This represents the function $A\,.\,\overline{B}+\overline{A}\,.\,B$, that is, the EXCLUSIVE-OR function, the map containing 1s in cells which are not adjacent to one another. These cells are defined by input conditions 10 and 01 and, as explained above, are not adjacent since their binary representations differ in more than one bit.

## A 3-variable Map

A 3-variable map is shown in figure 7.13 and, since there are $2^3 = 8$ possible combinations of the input variables, there are eight cells in the Karnaugh map. Once again, each variable must be represented in one-half of the total number of cells, so that variable $A$ is represented in four cells and $\overline{A}$ in the remaining four. Variable $B$ is also represented in four cells, two of which are associated with variable $A$ and two with $\overline{A}$; variable $\overline{B}$ also links with two cells containing $A$ and two containing $\overline{A}$. In this way, all possible combinations of variables $A$ and $B$ are obtained. Similarly variable $C$ links with both $A$ and $B$ so that all possible combinations of $A$, $B$ and $C$ are generated.

The binary code groups locating the cells are shown in figure 7.13 and, once more, we note that adjacent cells differ by only one bit in their respective code groups. An interesting feature of this map is the fact that the cells at the extreme left-hand side and the extreme right-hand side of each row are *adjacent cells* according to our definition. For example, the cells at the extremes of the upper row are the cells 000 and 100, which differ only in the left-hand digit; cells 001 and 101 in the bottom row differ only in the left-hand digit. Since we have *side-to-side adjacency*, we can 'bend' the map to form a continuous cylinder with the ends forming the 'seam' of the cylinder.

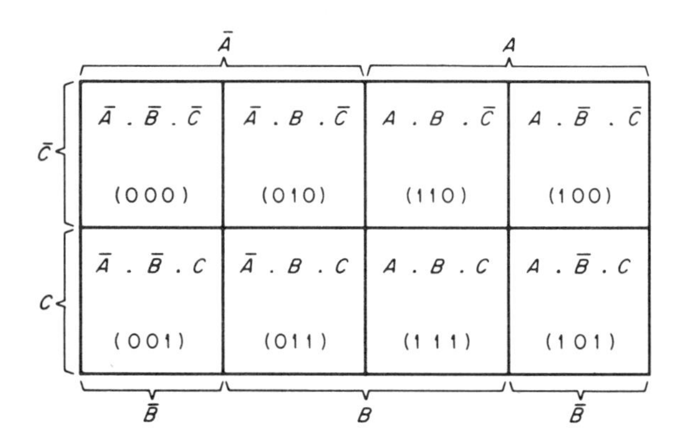


Figure 7.13 A three-variable Karnaugh map

Examples of function mapping on 3-variable maps are shown in figure 7.14. The function mapped in figure 7.14a contains 1s in non-adjacent cells and cannot, therefore, be simplified. The expression defined by this map is

$$f = A\,.\,\overline{B}\,.\,C+\overline{A}\,.\,B\,.\,\overline{C}$$

Figure 7.14b contains two pairs of adjacent cells. Let us first deal with the lower row. The cells grouped here are

$$A\,.\,B\,.\,C+\overline{A}\,.\,B\,.\,C = (A+\overline{A})\,.\,B\,.\,C = B\,.\,C$$

The equation defining this group can readily be deduced from the Karnaugh map by noting that the two centre cells in the bottom row represent the *total* number of cells covered by the *intersection* of variables $B$ and $C$. Similarly, the two cells containing 1s in the upper row have side-to-side adjacency and form the total number of cells represented by the intersection of variables $\overline{B}$ and $\overline{C}$. This fact is verified in the following

$$A\,.\overline{B}\,.\,\overline{C}+\overline{A}\,.\,\overline{B}\,.\,\overline{C} = (A+\overline{A})\,.\,\overline{B}\,.\,\overline{C} = \overline{B}\,.\,\overline{C}$$

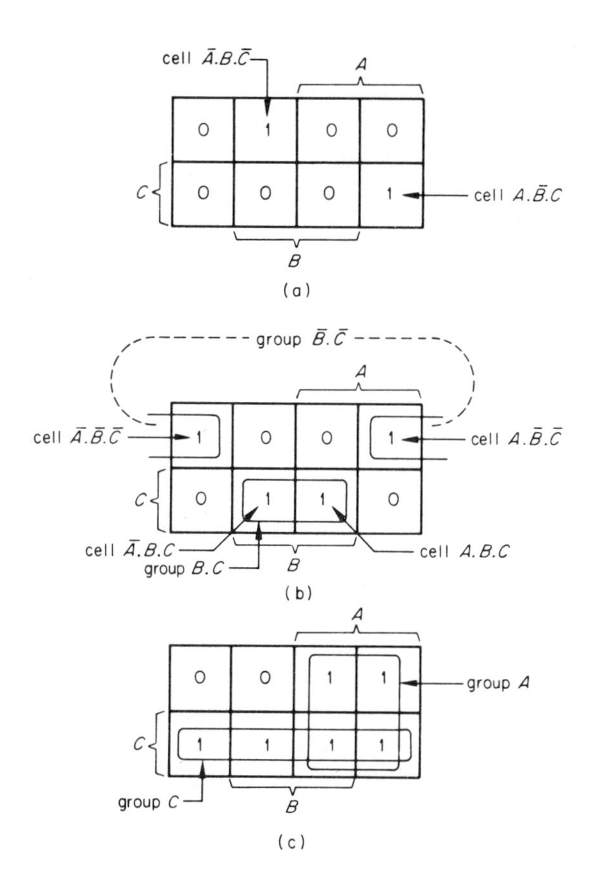


Figure 7.14 Illustrating adjacent cells on a three-variable map

Hence the function mapped in figure 7.14b is

$$f = B\,.\,C + \bar{B}\,.\,\bar{C}$$

Note that the variable $A$ is missing from the above expression, hence variable $A$ is *redundant* in this problem.

Two methods of grouping four adjacent cells are shown in figure 7.14c. Let us consider the right-hand group, which defines the logical expression

$$\begin{aligned} A\,.\,B\,.\,C + A\,.\,B\,.\,\bar{C} + A\,.\,\bar{B}\,.\,\bar{C} + A\,.\,\bar{B}\,.\,C & \\ = A\,.\,(B\,.\,C + B\,.\,\bar{C} + \bar{B}\,.\,\bar{C} + \bar{B}\,.\,C) & \\ = A\,.\,(B\,.\,(C + \bar{C}) + \bar{B}\,.\,(\bar{C} + C)) & \\ = A\,.\,(B + \bar{B}) = A & \end{aligned}$$

This result can be obtained directly from the Karnaugh map, since the four cells listed above are the total number of cells defined by variable $A$ on the map. It can also be seen that the four cells in the lower row represent variable $C$, this fact can be verified by a similar argument to that above. The logical function defined by the map in figure 7.14c is, therefore

$$f = A + C$$

It can now be noted that *the larger the number of cells grouped on the Karnaugh map, the simpler is the final logical expression.*

If three adjacent cells are to be grouped together on the map, then they may be grouped as two separate pairs in the manner outlined in figure 7.12c.

## A 4-variable Map

A 4-variable map containing $2^4 = 16$ cells is shown in figure 7.15a, an example of a 4-variable map is shown in figure 7.15b in which the function mapped is

$$f = \bar{A} + B\,.\,\bar{C}\,.\,D + \bar{B}\,.\,C\,.\,D + A\,.\,\bar{B}\,.\,\bar{D}$$

The eight cells which are grouped on the left of figure 7.15b collectively represent the term $\bar{A}$, while the term $B\,.\,\bar{C}\,.\,D$ is represented by the pair of cells grouped together in the centre of the map. An example of side-to-side adjacency is illustrated by the group $\bar{B}\,.\,C\,.\,D$.

Another facet of adjacency on Karnaugh maps illustrated by 4-variable maps is top-to-bottom adjacency. When we inspect the code groups at the head and foot of each column in figure 7.15a, we find that they satisfy the conditions associated with adjacent cells. That is to say, they differ in only one bit and we can bend the top of the map over to touch the bottom to form a cylinder. We have

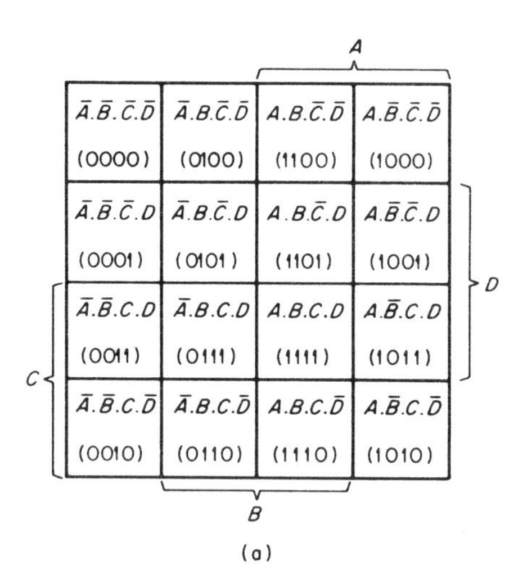


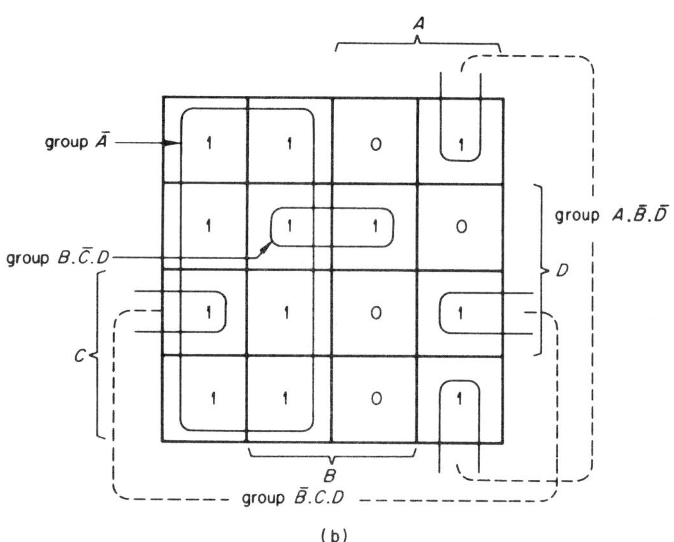


Figure 7.15 (a) A four-variable map and (b) the map of the function $f = \bar{A} + B.\bar{C}.D + \bar{B}.C.D + A.\bar{B}.\bar{D}$

already shown that the ends are adjacent, so that the Karnaugh map can be regarded as being a toroid or continuous cylinder, in the manner of an inner tube of a tyre. In this way, we see that the four corner cells in figure 7.15a form an adjacent group $\bar{B}.\bar{D}$ on the toroidal map. An example of top-to-bottom adjacency is illustrated in the group $A.\bar{B}.\bar{D}$.

The mapping technique can be extended to deal with more than four variables.

## 7.8 LOGIC CIRCUIT DESIGN USING DATA SELECTOR LOGIC (MULTIPLEXER LOGIC)

The design procedures outlined above are those to be adopted when considerable time is available, allowing many variations on a theme to be investigated. Unfortunately, designers' time is expensive and to reduce design time, many special ICs are available which allow complex networks to be replaced by one or more ICs, reducing the design time involved to a matter of minutes. One such group of devices is known as *data selectors* or *multiplexers*.

The basis of a 1-out-of-4 data selector is shown in figure 7.16. The circuit shown uses a logic network together with four transmission gates TG0 to TG3 (see also section 5.17). TTL versions of data selector logic are also available. The circuit shown could be housed in a 14-pin DIP, and would allow any one of four logic input signals ($I_0$ to $I_3$) to be selected using two address lines or 'select' lines $S_1$ and $S_2$. The output appears at terminal $Y$. The two 'select' lines allow us to select $2^2 = 4$ input lines. The data selector also has an inhibit ($I$) line; when this line is held at logic '0' it allows the data selector to function normally. For the moment we will assume that $I = 0$ so that $\bar{I} = 1$.

The transmission gate which is to be switched on is selected by the decoder circuit containing gates G0–G3. When the states of the selector lines are $S_1 = 0$, $S_2 = 0$, the output from G0 = 1 and that from gates G1, G2 and G3 are all logic '0'. These signals cause TG0 to be in its low-resistance state and TG1, TG2 and TG3 to be in a high-resistance state. Consequently the state of input $I_0$ is transmitted to the output terminal, and inputs $I_1$, $I_2$ and $I_3$ are isolated from it and from one another. The process of selecting a particular input line is known as *addressing* that line.

When $S_1 = 1$ and $S_2 = 0$, the output from G1 = 1 and the

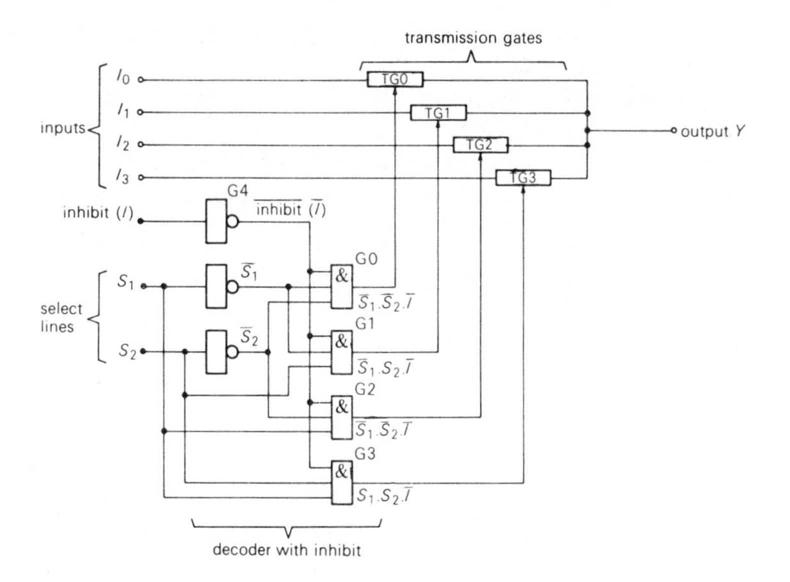


Figure 7.16 The basic circuit of a 1-out-of-4 data selector or multiplexer

outputs from G0, G2 and G3 are '0'. This condition results in TG1 being in its low-resistance state, and TG0, TG2 and TG3 being in a high-resistance state. The net result is that $I_1$ is connected to the output terminal.

When $S_1 = 0$, $S_2 = 1$, the state of $I_2$ appears at the output terminal, and when $S_1 = 1$, $S_2 = 1$, the logic state of $I_3$ is transmitted to the output.

Thus, altering the binary code applied to the 'select' lines or 'address' lines causes each input in its turn to be connected to the output.

The circuit shown is also known as a *multiplexer* since it allows multiple inputs to be transmitted sequentially over the same output wire (or busbar). The CMOS transmission gates in figure 7.16 allow information to be transmitted either from the input terminal to the output, or from the output to the input; such a circuit can be described as a multiplexer/demultiplexer. In the case of a TTL IC, each transmission gate would be replaced by an AND–OR network which permits data to be transmitted from the input to the output; for this reason the TTL version is described as a data selector/multiplexer.

The signal on the inhibit line controls the operation as follows. When $I = 0$ the output from G4 is $\bar{I} = 1$. This signal 'enables' gates G0–G3 to function normally. When the inhibit line is activated by a logic '1' signal, then $\bar{I} = 0$; this inhibits the operation of G0–G3, inclusive, so that all the transmission gates are in their high resistance state. Under this condition output $Y$ is isolated from all the input lines.

We will now consider the use of a 1-out-of-8 data selector/multiplexer* in solving the logic problem in section 7.2.

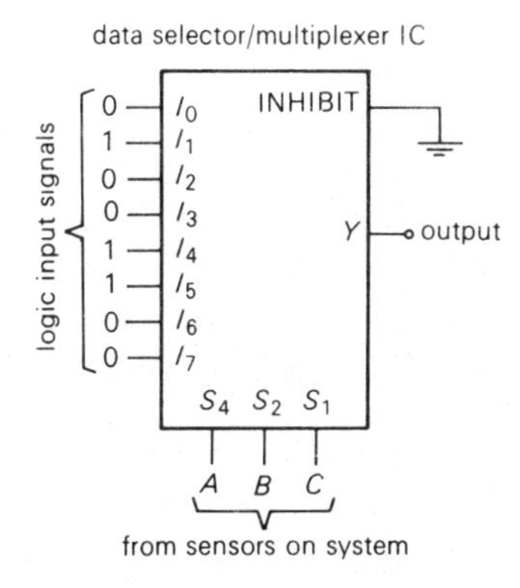


Figure 7.17 Using a data selector IC to solve the problem in example 7.2

The operational connections are shown in figure 7.17. Devices of this kind are available in a 16-pin DIP, and are MSI ICs. The logic signals connected to the input pins ($I_0$–$I_7$) of the IC correspond to the logic values given in the output column of the truth table (table 7.1). For convenience, these are listed below.

$$I_0 = 0,\ I_1 = 1,\ I_2 = 0,\ I_3 = 0,\ I_4 = 1,\ I_5 = 1,\ I_6 = 0,\ I_7 = 0$$

As explained earlier, the inhibit signal must be logic '0' for normal

* A CMOS IC fulfilling this function is the 4051, and a TTL IC is the 74151.

operation. This IC allows any one of the eight input lines to be selected using address lines $S_1$, $S_2$ and $S_4$, in which $S_1$ is the *least significant bit* (LSB) of the address code and $S_4$ is the *most significant bit* (MSB). The decoder part of the IC decodes the address of the input line in binary form with $S_4$ having a decimal 'weight' of 4, $S_2$ having a weight of 2 and $S_1$ having a weight of 1. Thus if $S_4 = 1$, $S_2 = 0$, $S_1 = 1$ (equivalent to decimal 5), the IC causes the signal applied to input $I_5$ to appear at output $Y$.

When the signals from sensors ABC have the value 000, the IC causes the logic state of $I_0$ ($= 0$) to appear at output $Y$. When the signals $ABC$ applied to the select pins have the value 001, the IC causes input $I_1$ ($= 1$) to appear at output $Y$, and so on. Hence the arrangement in figure 7.17 is equivalent to the logic network in figure 7.1. An advantage of a solution of this kind is that only a single chip is required to solve a complex problem. Moreover, the connections can be changed to accommodate alterations in the truth table with varying requirements of the system.

Although the IC described has only three address lines, it can be used to deal with four sets of address signals using a method sometimes described as the *folded method of solution*. Consider the truth table in table 7.4.

Here we 'fold' the truth table in half so that the first eight combinations (for which $A = 0$) are written down by the side of the second eight (for which $A = 1$).

**Table 7.4**

| Inputs | | | | Output | Inputs | | | | Output |
|---|---|---|---|---|---|---|---|---|---|
| $A$ | $B$ | $C$ | $D$ | $Y$ | $A$ | $B$ | $C$ | $D$ | $Y$ |
| 0 | 0 | 0 | 0 | 1 | 1 | 0 | 0 | 0 | 1 |
| 0 | 0 | 0 | 1 | 0 | 1 | 0 | 0 | 1 | 0 |
| 0 | 0 | 1 | 0 | 0 | 1 | 0 | 1 | 0 | 1 |
| 0 | 0 | 1 | 1 | 1 | 1 | 0 | 1 | 1 | 0 |
| 0 | 1 | 0 | 0 | 1 | 1 | 1 | 0 | 0 | 0 |
| 0 | 1 | 0 | 1 | 1 | 1 | 1 | 0 | 1 | 1 |
| 0 | 1 | 1 | 0 | 1 | 1 | 1 | 1 | 0 | 0 |
| 0 | 1 | 1 | 1 | 0 | 1 | 1 | 1 | 1 | 0 |

Putting aside input $A$ for the moment, the truth table is then re-written, but in this case we look at *pairs* of output values corresponding to identical sets of values of $B$, $C$ and $D$. The resulting truth table is given in table 7.5; an explanation of how it is obtained follows.

**Table 7.5**

| Inputs | | | Output |
|---|---|---|---|
| $B$ | $C$ | $D$ | $Y$ |
| 0 | 0 | 0 | $1 = I_0$ |
| 0 | 0 | 1 | $0 = I_1$ |
| 0 | 1 | 0 | $A = I_2$ |
| 0 | 1 | 1 | $\overline{A} = I_3$ |
| 1 | 0 | 0 | $\overline{A} = I_4$ |
| 1 | 0 | 1 | $1 = I_5$ |
| 1 | 1 | 0 | $\overline{A} = I_6$ |
| 1 | 1 | 1 | $0 = I_7$ |

Starting with the first set of values in table 7.4, we see that for the address combination (0)000 (the state of $A$ being given in parenthesis) the output is '1', and for (1)000 the output is also '1'; thus a logic '1' is applied to input $I_0$ of the IC, since $Y = 1$ whatever the value of signal $A$. Hence a '1' is written down in the 'output' column of table 7.5. For the address combination (0)001 the output is '0', and for the combination (1)001 it is also '0'; a logic '0' is therefore applied to input $I_1$ of the IC. In the case of the combination (0)010 the output is '0', and for (1)010 the output is '1'; that is $Y = A$ for the combination $BCD = 010$, so that we connect signal $A$ to input $I_2$ of the IC. When the combination is (0)011 the output is '1', and when it is (1)011 it is '0'; that is $Y = \overline{A}$ for the combination $BCD = 011$, hence $\overline{A}$ is connected to $I_3$ of the IC. It is left as an exercise for you to verify the remainder of table 7.5. The resulting connections to the IC which satisfy table 7.4 are shown in figure 7.18.

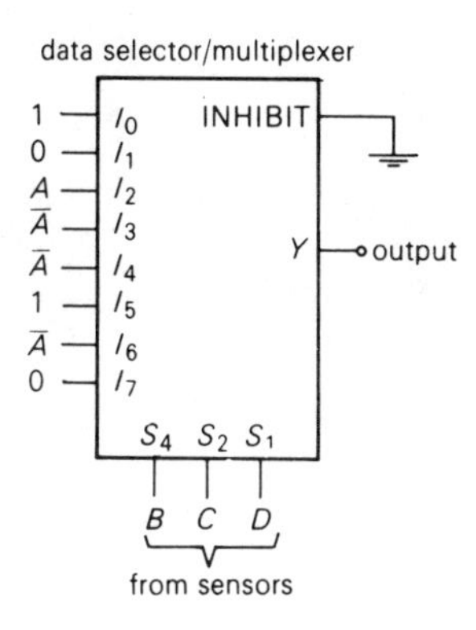


Figure 7.18 Data selector logic solution for table 7.4

## PROBLEMS

**7.1** Design a minimal logic network containing AND, OR and NOT gates which satisfies table 7.6.

**Table 7.6**

| *Inputs* A | B | C | *Output* f |
|---|---|---|---|
| 0 | 0 | 0 | 0 |
| 0 | 0 | 1 | 1 |
| 0 | 1 | 0 | 1 |
| 0 | 1 | 1 | 1 |
| 1 | 0 | 0 | 1 |
| 1 | 0 | 1 | 0 |
| 1 | 1 | 0 | 1 |
| 1 | 1 | 1 | 1 |

**7.2** Design a NAND network which satisfies table 7.6.

**7.3** Design a NOR network which satisfies table 7.6.

**7.4** Design a majority voting system, whose specification is given below, which has inputs from three microprocessors shown in figure 7.19. The output from the majority voting system must be logic '0' if two or more of the signals from processors A, B and C are '0', and its output must be logic '1' if two or more of the signals from the processors are logic '1'.

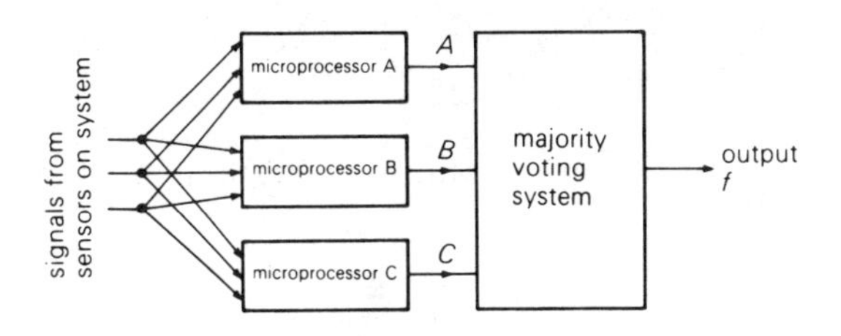


Figure 7.19

**7.5** Redesign the system in problem 7.4 using only NOR gates.

**7.6** (a) What input conditions at $A$ and $B$ in figure 7.20 give $f = 0$? (b) What input conditions give $f = 1$? Give reasons for your solutions.

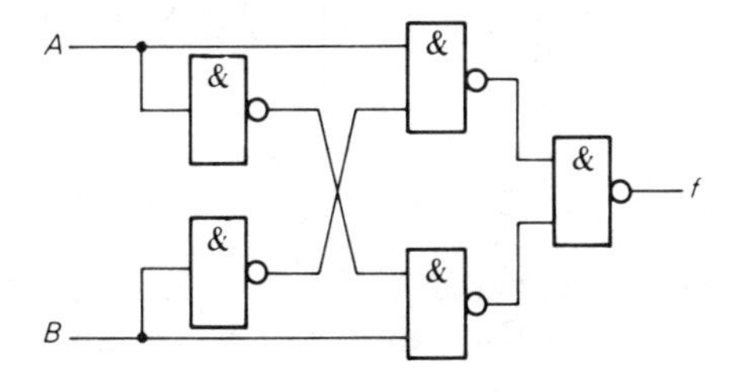


Figure 7.20

**7.7** Complete the missing details in table 7.7.

**7.8** Design a network containing only NOR gates which generates the logic function $\overline{\overline{A} \, . \, (B + \overline{C})}$.

**Table 7.7**

| Inputs A | Inputs B | Output $f = \overline{A} + B$ |
|---|---|---|
| 0 | 0 | 1 |
| 0 | 1 | 1 |
| 1 | 1 | 1 |

**7.9** Design a system using a 1-out-of-8 data selector IC and a NOT gate to satisfy table 7.8.

**Table 7.8**

| A | B | C | D | Output | A | B | C | D | Output |
|---|---|---|---|---|---|---|---|---|---|
| 0 | 0 | 0 | 0 | 1 | 1 | 0 | 0 | 0 | 0 |
| 0 | 0 | 0 | 1 | 1 | 1 | 0 | 0 | 1 | 1 |
| 0 | 0 | 1 | 0 | 1 | 1 | 0 | 1 | 0 | 1 |
| 0 | 0 | 1 | 1 | 0 | 1 | 0 | 1 | 1 | 1 |
| 0 | 1 | 0 | 0 | 1 | 1 | 1 | 0 | 0 | 0 |
| 0 | 1 | 0 | 1 | 0 | 1 | 1 | 0 | 1 | 1 |
| 0 | 1 | 1 | 0 | 0 | 1 | 1 | 1 | 0 | 0 |
| 0 | 1 | 1 | 1 | 1 | 1 | 1 | 1 | 1 | 1 |

**7.10** The waveforms in figure 7.21 are applied to the inputs of an EXCLUSIVE-OR gate. Draw the output waveform from the gate.

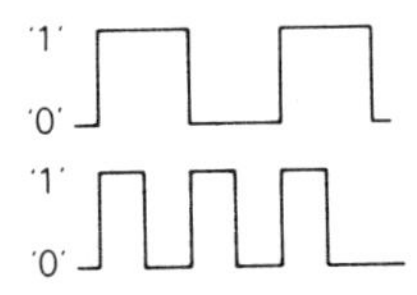


Figure 7.21

# 8 Flip–Flops

## 8.1 STATIC AND DYNAMIC MEMORIES

Sequential logic systems include such networks as counters and shift registers, and require the use of memory elements known as *flip–flops* or *bistables* to record the state of the circuit at a particular instant of time.

Semiconductor memory circuits fall into two broad categories: *static memories* and *dynamic memories*. Static memories retain the stored information indefinitely so long as the power supply is maintained; both bipolar and MOS devices are used in static memory elements. The most popular types of static memory elements are *set–reset* (*S–R*) *flip–flops*, *J–K flip–flops*, *trigger* (*T*) *flip–flops* and *D flip–flops*. The J–K element is the most versatile of these, since it can be used to generate all other types of memory function.

Dynamic memories depend for their operation on the ability of the gate capacitance of MOS devices to retain their charge for relatively long periods of time (long, that is, compared with the time for a complete cycle of operations within the system, which may only be a few hundered μs in a computer system). Dynamic memories are dealt with in section 8.10.

## 8.2 THE S–R FLIP–FLOP

A basic S–R flip–flop using cross-connected NOR gates is shown in figure 8.1a. The output from the circuit is designated the letter $Q$, its logical complement being available at terminal $\overline{Q}$. A logical '1' signal applied to the S-line causes $Q$ to be set to the logic '1' level irrespective of its previous state. We may therefore regard the S-line as the line which allows us to *set* the output to the '1' state. At the same instant of time, output $\overline{Q}$ becomes '0'. The application of a signal to the R-line causes $Q$ to be *reset* to '0' (that is, $\overline{Q}$ is set to '1'). A detailed description of the operation of the circuit follows.

Let us assume that $Q = 0$ initially (that is, $\overline{Q} = 1$), and that the signals applied to both input lines are zero. This state will be seen to be a stable operating condition, since the logic '1' output from G2 which is fed back to G1 holds the output of G1 at '0'. Both inputs to G2 are 0s, so that its output is '1'.

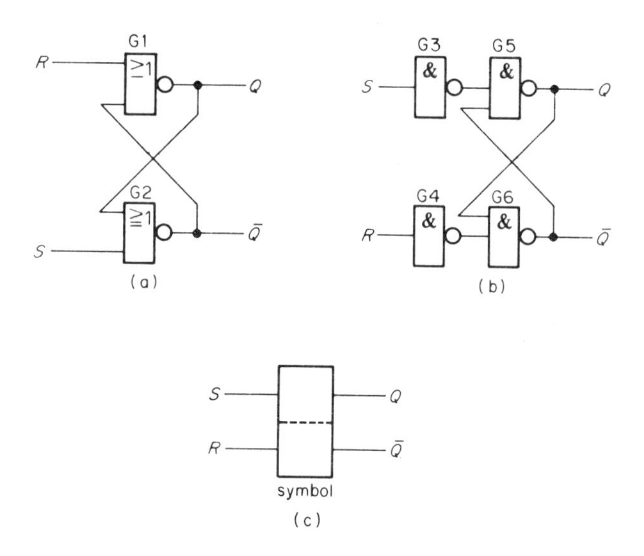


Figure 8.1 An S–R flip–flop or bistable using (a) NOR gates and (b) NAND gates. (c) Circuit symbol of the S–R flip–flop

The application of a logic '1' signal to the S-line causes the output of G2 to fall to zero and, since this is fed back to the input of G1, then the output of G1 rises to logic '1'. This is the second stable operating condition with $Q$ having been set to '1'. The signal applied to the S-line need only be applied momentarily, since the feedback connections between the gates cause the flip–flop to memorise the last instruction. The application of further pulses to the S-line has no further effect on the state of the circuit, since it has already been set to the '1' state.

It can be shown by a similar argument that an impulsive '1' signal applied to the R-line (with input $S = 0$) causes $Q$ to be reset to zero.

The application of '1' signals to both S and R lines simultaneously causes both outputs to fall to zero. The final state of the outputs when both input signals are removed simultaneously is indeterminate, as this depends on the relative switching speed of the two gates. *This operating condition is avoided in practice.*

Since we are concerned with a sequence of events which changes with time, we define the operation of the network in terms of its *sequential truth table*, which is given in table 8.1

**Table 8.1**

| $S$ | $R$ | $Q_{n+1}$ | *Comment* |
|---|---|---|---|
| 0 | 0 | $Q_n$ | no change |
| 0 | 1 | 0 | reset |
| 1 | 0 | 1 | set |
| 1 | 1 | ? | indeterminate |

In table 8.1, $Q_n$ is the state of the flip–flop output after $n$ operations have occurred, and $Q_{n+1}$ is the state of $Q$ after $n+1$ operations. Let us assume that, in each case, we have completed $n$ operations, and that the next step of input conditions are the $(n+1)$th set of input signals.

As stated above, when $S = R = 0$ the output remains unchanged whatever its previous value (it may have either been '1' or '0'). That is, $Q_{n+1} = Q_n$. The second row of the truth table corresponds to the reset condition ($R = 1$, $S = 0$), causing $Q$ to be zero after the operation. The third row of the truth table is the set operation ($S = 1$, $R = 0$), which causes $Q$ to be set to the '1' state. When $S = R = 1$, in the case of a NOR memory, both outputs are zero. This is described as an indeterminate or 'don't know' condition since, as outlined above, the final state of the output when the input signals are reduced to zero is indeterminate.

A NAND S–R flip–flop is shown in figure 8.1b, and includes invertors G3 and G4 in the input lines. The flip–flop itself comprises gates G5 and G6. The explanation of the operation of this circuit is left as an exercise for readers to test their skill on. The S–R flip–flop (either NOR or NAND) is represented by the symbol in figure 8.1c.

## 8.3 THE GATED S–R FLIP–FLOP

It is often convenient to control all the operations in a system by means of a common synchronising pulse or *clock pulse*. By this means it is possible to *gate* or to *clock* signals into the flip–flop at a precise time. One method of doing this is shown in figure 8.2.

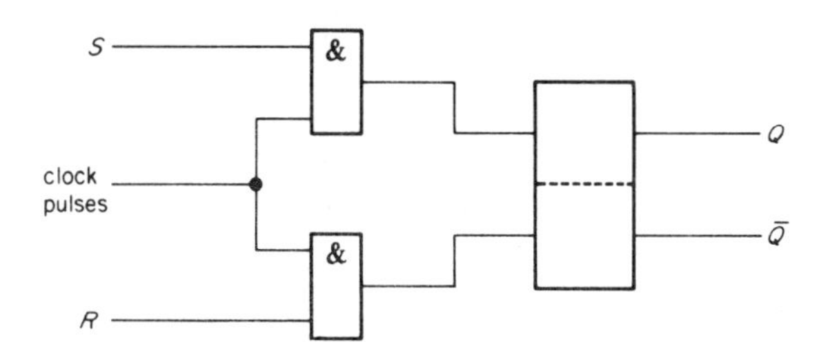


Figure 8.2 A gated S–R flip–flop

In the case of a NAND flip–flop of the type in figure 8.1b, no additional gates are required as the clock signal would be applied simultaneously to gates G3 and G4.

## 8.4 CONTACT BOUNCE ELIMINATORS

Electrical switches are often used as a means of applying signals to digital systems, but all conventional switches generate electrical noise which results from 'contact bounce'. If the system to which the signal is applied is a counter, then the system counts each noise pulse as though it were a genuine *on–off* signal. Two methods of eliminating the effects of contact bounce are shown in figure 8.3.

With an S–R flip–flop using NOR gates, the circuit in figure 8.3a

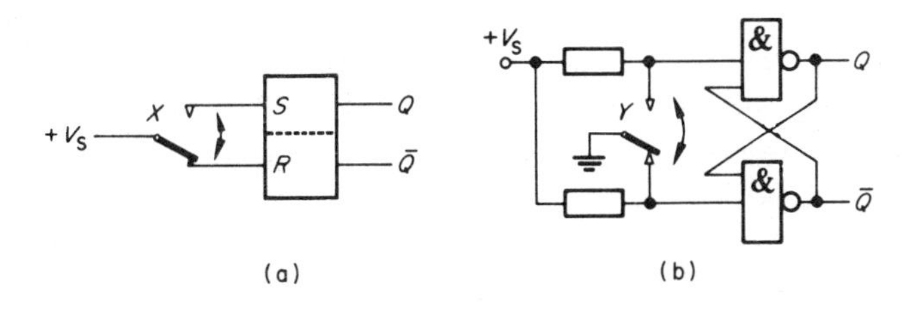


Figure 8.3 Contact bounce elimination circuits

is adopted since, in NOR networks, the gates are switched by logic '1' signals. When switch X is moved to the S position, output $Q$ instantaneously becomes '1' and does not alter no matter how many times the switch contacts bounce. Output $Q$ is used as the effective output from the switch.

In NAND circuits, logic '0' signals control the switching operation. The circuit in figure 8.3b is one which is frequently used for contact bounce elimination in NAND circuits. Switching the zero potential point to the input of one of the gates causes the output of that gate to be '1'.

## 8.5 MASTER–SLAVE FLIP–FLOPS

In the quest for higher operating speeds, various types of flip–flop were developed and the basis of modern flip–flops is the master–slave circuit.

The underlying principle of the master–slave flip–flop is shown in figure 8.4a, which contains two S–R flip–flops connected by synchronously operated switches S1 and S2. The switches are connected so that when S1 is open then S2 is closed, and vice versa. When the clock signal is '0', S1 is open and S2 is closed, so that the information stored in the master is transmitted to the slave. The waveform in figure 8.4b is that of the clock signal applied to the flip–flop, and the conditions outlined above occur in period A.

When the clock signal rises to the '1' level, corresponding to period B in figure 8.4b, S1 closes and S2 opens. At this instant of time, new data is transferred into the master, while the slave retains the previous data. When the clock signal falls to zero once more, period C in figure 8.4b, S1 opens and S2 closes. This isolates the master from the input and connects it to the slave, and the new data is transferred into the slave. Thus, *the new input data is gated to the output of the flip–flop at the trailing edge of the clock pulse.*

Since the operation of the master–slave flip–flop described here is controlled by the logic level of the clock pulse, it is sometimes described as a *level-triggered* flip–flop.

A block diagram of a *master–slave S–R flip–flop* is shown in figure 8.5, in which G1 and G2 are equivalent to S1 in figure 8.4a, and G3 and G4 are equivalent to S2. The invertor G5 provides the

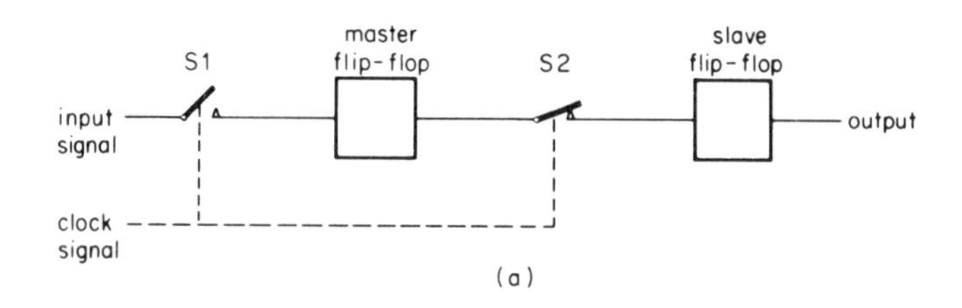


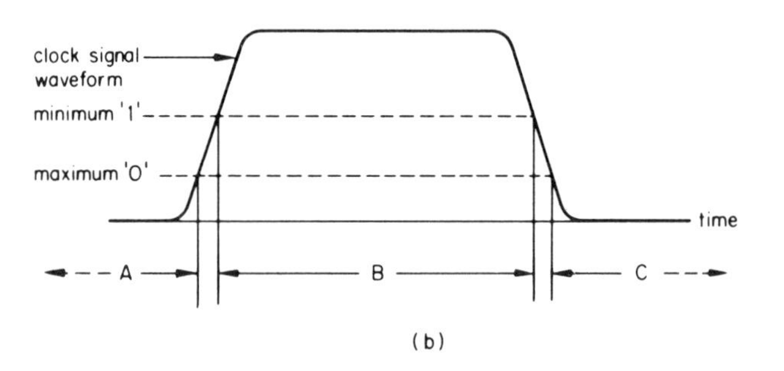


Figure 8.4 (a) General principle of a master–slave flip–flop and (b) its timing sequence

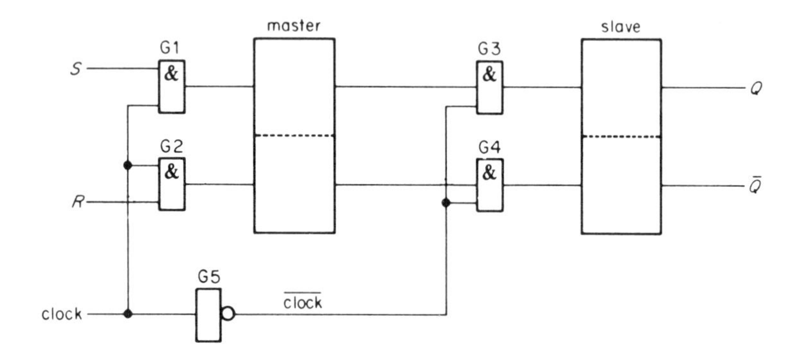


Figure 8.5 One form of master–slave S–R flip–flop

correct phase relationship between the two switches. The master and slave flip–flops in figure 8.5 are circuits of the type described in section 8.2.

TTL monolithic IC flip–flops can operate at clock frequencies of up to about 35 MHz.

## 8.6 THE MASTER–SLAVE J–K FLIP–FLOP

A basic form of master–slave J–K flip–flop is shown in figure 8.6. It is similar in structure to the master–slave S–R flip–flop, with the exception that the outputs are fed back as shown in figure 8.6.

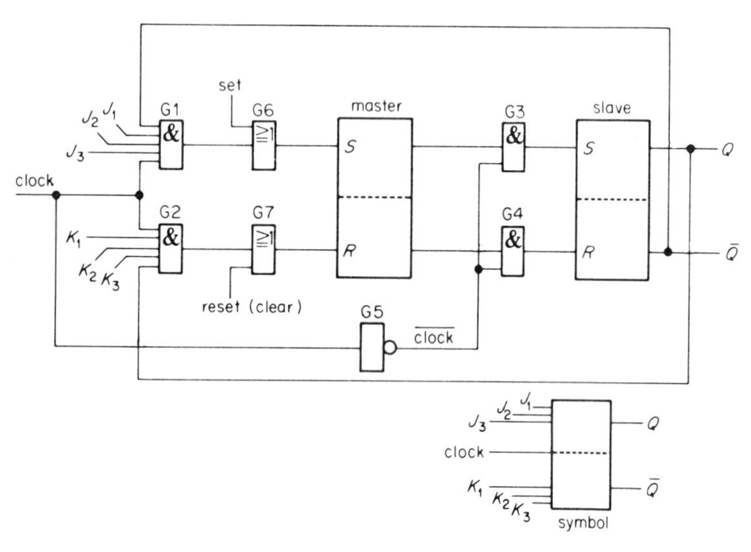


Figure 8.6 A J–K master–slave flip–flop

Other input lines are added to the circuit and, although they do not alter the operation of the circuit, they increase the versatility of the flip–flop. The addition of gates G6 and G7 allow the output either to be set to the '1' state or to be reset to the '0' state. In some flip–flops the reset line is known as the *clear* input. As before, gates G1 and G2 are equivalent to S1 in figure 8.4a and G3 and G4 are equivalent to S2 in the same figure. The sequential truth table for figure 8.6 is given in table 8.2.

**Table 8.2**

| $J$ | $K$ | $Q_{n+1}$ | *Comments* | |
|---|---|---|---|---|
| 0 | 0 | $Q_n$ | No change | S–R operation |
| 0 | 1 | 0 | reset | S–R operation |
| 1 | 0 | 1 | set | S–R operation |
| 1 | 1 | $\overline{Q}_n$ | toggle or trigger | |

Before going on to describe the operation of the circuit, we shall discuss the implications of the truth table. Comparing tables 8.1 and 8.2, we see that if we regard the J-input and the K-input lines as being equivalent to the S-input and R-input, respectively, of the S–R flip–flop, then for the first three groups of input conditions the S–R and J–K truth tables are equivalent to one another. That is, the S–R flip–flop may be replaced by the J–K flip–flop.

Also, when $J = K = 1$, the circuit acts as a *toggle flip–flop* or *trigger flip–flop* whose output changes state at the end of *each* clock pulse. That is, if $Q = 0$ initially, then at the end of the first clock pulse the output changes to '1', at the end of the second clock pulse it changes to '0' again, and the process is repeated indefinitely so long as $J = K = 1$. Flip–flops operating in the toggle or trigger mode are widely used in *asynchronous counting systems* (see chapter 10). The operation of the circuit in figure 8.6 is now described.

When $J = K = 0$, both G1 and G2 are inhibited and no signals can be applied to the master flip–flop. As a result, the output remains unchanged and $Q_{n+1} = Q_n$.

The operation for $J = 1$, $K = 0$ is considered in two parts, namely the operation when the initial value of $Q$ is '0' and when it is '1'. Let us consider the case when $Q = 0$. In this case $\overline{Q} = 1$, and when the clock signal is applied it causes G1 to be opened to allow the '1' signal to be applied to the set line of the master flip–flop. Gate G2 is inhibited not only by the '0' on the K-line but also by the '0' fed back from the $Q$ output. When the clock signal finally falls to zero, the '1' stored in the $Q$ output of the master stage is gated into the slave section of the flip–flop, causing $Q$ to change from '0' to '1'.

In the case when $Q$ is initially '1' ($\overline{Q} = 0$), the $\overline{Q}$ signal inhibits the operation of G1, and the states of the master and slave remain unaltered and the output remains at '1'.

The circuit operation for $J = 0$, $K = 1$ is similar to that described above, but for $J$ read $K$, for $K$ read $J$, for G1 read G2, for G2 read G1, for $Q$ read $\overline{Q}$, and for $\overline{Q}$ read $Q$.

For the input condition $J = K = 1$, then *either* G1 *or* G2 is opened when the clock signal is applied, the gate selected being dependent upon the signals fed back from the output. Suppose, initially, that $Q = 0$ and $\overline{Q} = 1$. These signals result in G1 being opened and G2 being inhibited, so that a '1' is fed from the J-line into the master flip–flop. At the end of the first clock pulse, this '1' signal is applied to the set-line of the slave, causing $Q$ to change from '0' to '1'. The feedback conditions have now changed so that a '1' is fed back to G2 and a '0' is applied to G1, thereby opening G2 and inhibiting G1. The next clock pulse causes a '1' to be applied to the reset-line of the master flip–flop. This, at the end of the clock pulse, causes output $Q$ to change from '1' to '0', thereby setting up conditions for a '1' to be fed to $Q$ during the following clock pulse period.

During the period of time that the clock signal is '0', gates G3 and G4 are opened, so that the output can either be set or reset (cleared) by the application of a signal to the set or reset-lines connected to G6 and G7, respectively.

## 8.7 THE TRIGGER (T) FLIP–FLOP

A flip–flop with a trigger or toggle operation is constructed merely by connecting the J and K-inputs to a logic '1' signal as shown in

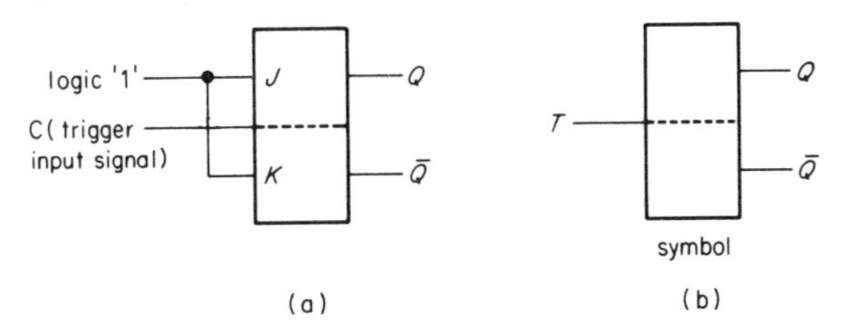


Figure 8.7 A trigger (T) flip–flop constructed from a J–K flip–flop

figure 8.7a. The circuit then functions in the manner described in the penultimate paragraph of section 8.6. A symbol frequently used for the T flip–flop is shown in figure 8.7b, the T-line being synonymous with the C-line in figure 8.7a.

## 8.8 THE D MASTER–SLAVE FLIP–FLOP

A basic block diagram of a D-type master–slave flip–flop is shown in figure 8.8, and consists of a J–K flip–flop whose K-input is driven via an invertor. The result of adding this invertor is that the signals applied to the J and K-inputs are *always* complementary, and the truth table of the D flip–flop corresponds to the second and third lines of table 8.2, which are listed in table 8.3.

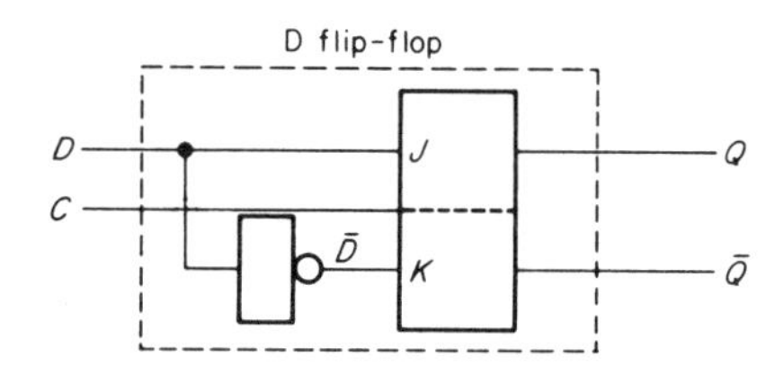


Figure 8.8 A delay (D) flip–flop constructed from a J–K flip–flop

**Table 8.3**

| $D$ | $Q_{n+1}$ | *Comment* |
|---|---|---|
| 0 | 0 | reset |
| 1 | 1 | set |

The D flip–flop, known as a *delay* flip–flop or as a *data* flip–flop, delays the transmission of data between the input and output by a time interval equal to one clock pulse. It is widely used as a data latching buffer element between a counting circuit and a digital read-out device.

The D flip–flop has the advantage over the J–K flip–flop that it only has one input line, resulting in simple interconnections between elements.

Toggle or trigger operation is obtained using the D flip–flop by feeding back the $\bar{Q}$ output as shown in figure 8.9, and by using the clock line as the T-line.

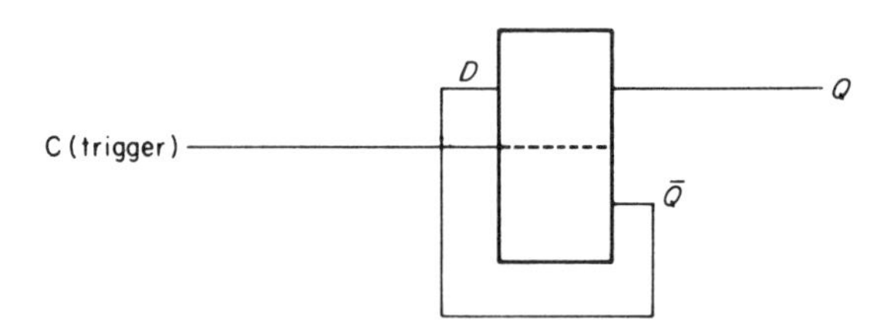


Figure 8.9 A T flip–flop constructed from a D flip–flop

## 8.9 EDGE-TRIGGERED FLIP–FLOPS

In the level-triggered master–slave flip–flops described above, the master flip–flop remains active (that is, it is connected to the input line) during the time that the clock signal is at the '1' level, and the data is finally transferred to the output when the clock signal falls to '0'.

In the range of flip–flops described as edge-triggered flip–flops, data is transferred to the output *on the incidence* of the leading edge of the clock pulse.

The data input must be applied to the input terminals for a minimum period of time known as the *set-up time* (typically 10 ns) prior to the clock pulse reaching its threshold value (typically 1.5 V in a TTL device). After the clock pulse has reached this value, the input information must remain for a period of time known as the *hold time*, which is typically 0–5 ns. After this point the input information is 'locked-out', and has no further effect during the clock cycle.

## 8.10 DYNAMIC MEMORIES

Dynamic memories depend for their operation on the ability of the parasitic capacitance of the gate of a MOST to retain an electrical charge. The simplest, and most popular form of MOS dynamic memory is the three transistor circuit of figure 8.10.

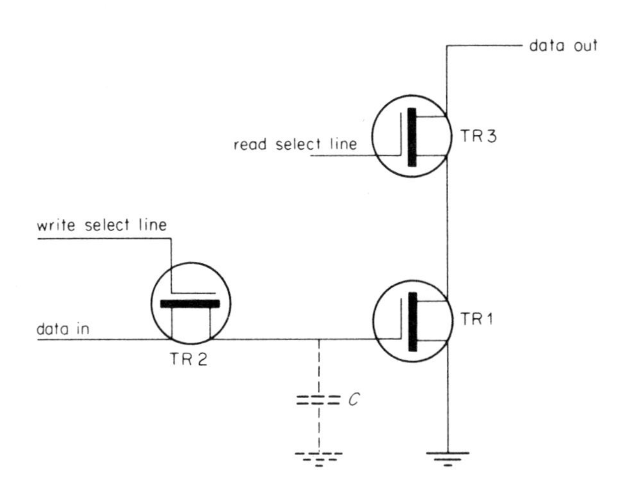


Figure 8.10 A basic form of dynamic memory

Transistor TR1 is used as the active device which stores the information, the parasitic gate capacitance C being the storage element. A logic '1' is 'written' into the memory by charging capacitor C by applying a signal to the gate of TR2 via the *write select* line, and then by charging C by energising the *data* line. After a period of time the charge on C decays a little, and the information is *refreshed* by writing the data in once more. The information is, typically, refreshed every 2 ms or so. If we wish to discharge C, the data line is held at zero potential while the write select line is activated.

To *read* the data stored in the memory, TR3 is turned *on* by the application of a signal to the *read select* line, the state of the cell being detected by sensing the current flowing in the *data out* line. The *access time* to the information stored is, typically, 150 ns.

## PROBLEMS

**8.1** The waveforms in figure 8.11a are applied to the circuit in figure 8.11b. Draw the waveform appearing at output $X$ if its value is initially logic '1'.

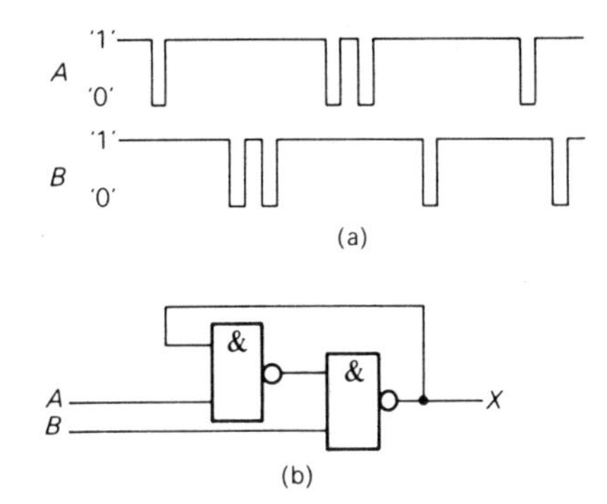


Figure 8.11

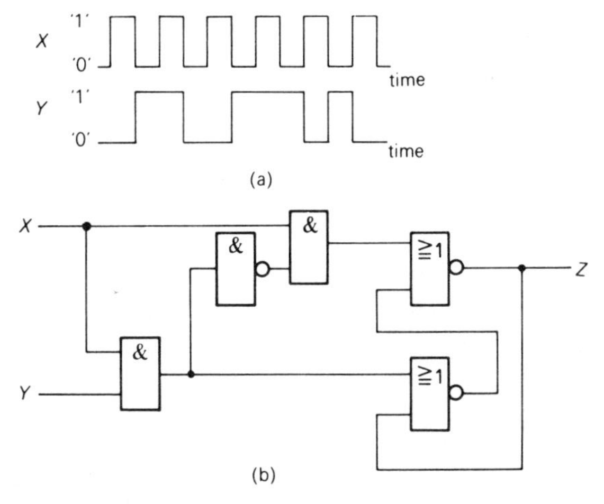


Figure 8.12

**8.2** What are the important advantages of the J–K master–slave flip–flop over the conventional S–R bistable circuit consisting of two cross-connected NOR gates.

**8.3** Draw a logic block diagram of a J–K master–slave flip–flop using only NAND gates. Show how 'set' and 'clear' lines can be added.

**8.4** The waveforms in figure 8.12a are applied to the circuit in figure 8.12b. Draw the waveform which appears at output $Z$.

# 9 Binary Codes and Arithmetic Processes

## 9.1 THE BINARY SYSTEM

Just as we express a decimal number as a series of multiples of powers of ten, it is also possible to express the same number as a series of multiples of powers of 2. For example, the decimal number 14 is represented in decimal form as

$$(1 \times 10^1) + (4 \times 10^0)$$

The same number expressed in binary form is

$$(1 \times 2^3) + (1 \times 2^2) + (1 \times 2^1) + (0 \times 2^0)$$

or, expressed in *binary-coded decimal* form (see also section 9.9) the decimal number 14 is $8+4+2+0 = 14$.

The total number of digits used as the basis of the numbering system is known as the *radix* of the system. The decimal (or denary) system uses the ten digits 0, 1, 2, . . . , 7, 8 and 9, and has a radix of ten; the binary system uses the digits 0 and 1 and has a radix of two. The relationship between the two systems for the first sixteen values (zero to decimal 15) is given in table 9.1, the binary sequence shown being known as the *natural binary code* or *pure binary code*. Each term in the pure binary code is given an equivalent decimal value or *weight*, which commences to the immediate left of the *binary point* (which is the binary equivalent of the decimal point) with unity value, and doubles its decimal value for each succeeding digit, that is, commencing at the binary point the weights increase in the sequence 1, 2, 4, 8, 16, 32, etc. A feature of the binary system is that each digit can assume only one of two possible values, corresponding to the '0' and '1' logic levels. Each *bi*nary digi*t* is referred to as a *bit*, and a four-bit code group of the type in table 9.1 can deal with up to $2^4 = 16$ different values.

Note that the binary value 10 (decimal 2) is pronounced 'one, zero' and not 'ten'; the binary value 101 (decimal 5) is pronounced 'one, zero, one', etc.

Two other codes frequently used in conjunction with microprocessors and computers are the *octal code* (base 8) and the *hexadecimal code* (base 16). The first twenty numbers in these codes are given in table 9.2. Note that in the hexadecimal code

**Table 9.1**

| | *Decimal* | | *Binary* | | | |
|---|---|---|---|---|---|---|
| *Decimal 'weight'* | $10^1$ (10) | $10^0$ (1) | $2^3$ (8) | $2^2$ (4) | $2^1$ (2) | $2^0$ (1) |
| | 0 | 0 | 0 | 0 | 0 | 0 |
| | 0 | 1 | 0 | 0 | 0 | 1 |
| | 0 | 2 | 0 | 0 | 1 | 0 |
| | 0 | 3 | 0 | 0 | 1 | 1 |
| | 0 | 4 | 0 | 1 | 0 | 0 |
| | 0 | 5 | 0 | 1 | 0 | 1 |
| | 0 | 6 | 0 | 1 | 1 | 0 |
| | 0 | 7 | 0 | 1 | 1 | 1 |
| | 0 | 8 | 1 | 0 | 0 | 0 |
| | 0 | 9 | 1 | 0 | 0 | 1 |
| | 1 | 0 | 1 | 0 | 1 | 0 |
| | 1 | 1 | 1 | 0 | 1 | 1 |
| | 1 | 2 | 1 | 1 | 0 | 0 |
| | 1 | 3 | 1 | 1 | 0 | 1 |
| | 1 | 4 | 1 | 1 | 1 | 0 |
| | 1 | 5 | 1 | 1 | 1 | 1 |

(sometimes abbreviated to 'hex code') the letters a to f are used to represent the decimal values 10 to 15.

## 9.2 FRACTIONAL NUMBERS

As with decimal numbers, binary numbers which are less than unity in value are represented by a series of multiples of powers of 2, the powers to which the radix is raised having a negative sign, as follows.

| binary value | decimal value |
|---|---|
| $1 \times 2^{-1} = 0.1$ | 0.5 |
| $1 \times 2^{-2} = 0.01$ | 0.25 |
| $1 \times 2^{-3} = 0.001$ | 0.125 |
| $1 \times 2^{-4} = 0.0001$ | 0.0625 |

**Table 9.2**

| *Decimal value* | *Octal* | *Hexadecimal* |
|---|---|---|
| 0 | 0 | 0 |
| 1 | 1 | 1 |
| 2 | 2 | 2 |
| 3 | 3 | 3 |
| 4 | 4 | 4 |
| 5 | 5 | 5 |
| 6 | 6 | 6 |
| 7 | 7 | 7 |
| 8 | 10 | 8 |
| 9 | 11 | 9 |
| 10 | 12 | a |
| 11 | 13 | b |
| 12 | 14 | c |
| 13 | 15 | d |
| 14 | 16 | e |
| 15 | 17 | f |
| 16 | 20 | 10 |
| 17 | 21 | 11 |
| 18 | 22 | 12 |
| 19 | 23 | 13 |
| 20 | 24 | 14 |

Thus, the decimal number 6.625 is represented in binary form as 110.101, that is, $4+2+0+0.5+0+0.125$.

## 9.3 BINARY ADDITION

Whatever the radix of the numbering system, the same mathematical processes are involved in the addition of two numbers. That is, if the sum of two numbers is less than the radix, we simply write down the sum. For example, if we add the decimal numbers 4 and 5 together, we write down the sum as 9. If the sum is greater than the radix, then we record the amount by which the sum is greater than the radix, and carry a '1' forward to the next higher column of the addition. It is important to note that when we add *two* decimal values together, the 'carry' is *either* zero *or* it is unity, and is never any other value. Hence, if we add the decimal numbers 9 and 8 together, we record a 7 and carry 1 to the next higher column of the addition. In the above case, the carry digit is known as the *carry out*, $C_O$. This is carried forward to the next addition, when it is known as the *carry-in*, $C_I$.

Thus, the addition process can be regarded as consisting of two steps. In the first part we add the two numbers, which are known as the *addend* and the *augend*, and generate a *sum* and a *carry*. We then add to the sum the carry-in generated by the previous calculation. The two parts are described as 'half-additions', the net result being a full addition. Electronic adding circuits use two half-adders (see section 9.4) to form a full adder. The complete addition process for binary numbers is illustrated in the following.

If we wish to add together in binary the two numbers which are equivalent to decimal 11 and 14, we proceed as follows

| decimal | binary | comment |
|---|---|---|
| 11 | 1011 | augend |
| 14 | 1110 | addend |
| | 0101 | first partial sum |
| | 1010 | first half-addition carry |
| | 10001 | second partial sum |
| | 0100 | second half-addition carry |
| 25 | 11001 | sum |

We shall now consider the process step-by-step. In the $2^0$ column we must first add $1 + 0 = 1$, giving a carry of zero. For convenience, this carry-out (0) is shifted one place to the left so that it will be in the correct position to be 'carried-in' to the $2^1$ addition. In the $2^1$ column we have the addition $1 + 1$ in binary. However, we cannot write down the number 2 as the sum, since the binary equivalent of 2 is 10 (see table 9.1). Thus $1 + 1 = 0$, carry 1. Once more the carry-out, this time a '1', is shifted to the left to become the carry-in to the $2^2$ addition. This process is continued until we have formed the partial sum of the addend and the augend, together with the associated carry bits.

The first partial sum and the first half-addition carry are added together to give a second partial sum and a second carry. The addition of these two binary groups gives the correct sum of the two numbers.

## 9.4 ADDITION NETWORKS

In the above it was shown that the addition process could be treated in two parts, that is, as a combination of two half-adder stages. Let us consider the truth table for the first half-addition of the addend and augend, $A$ and $B$, which is shown in table 9.3.

**Table 9.3**

| *Inputs* | | *Sum* | *Carry* |
|---|---|---|---|
| $A$ | $B$ | $S$ | $C$ |
| 0 | 0 | 0 | 0 |
| 0 | 1 | 1 | 0 |
| 1 | 0 | 1 | 0 |
| 1 | 1 | 0 | 1 |

Inspecting the relationship which exists between inputs $A$ and $B$ and the sum $S$, we see that

$$S = A \cdot \overline{B} + \overline{A} \cdot B$$

That is, a *sum* is generated when $A$ is *not equivalent* to $B$. All we need, therefore, to generate the sum of two binary numbers is a *not-equivalent* gate.

Inspecting the relationship which exists between the inputs and the carry signal, $C$, we find that

$$C = A \cdot B$$

Combining these relationships in the form of a logic circuit, the network in figure 9.1 generates both the sum and the carry associated with the addition. Alternatively, we could use the circuit developed in chapter 7 (figure 7.7c) which generates both the sum and the carry functions.

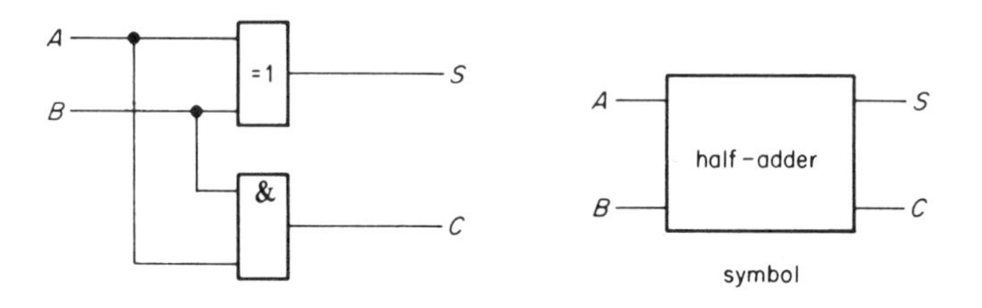


Figure 9.1 A half-adder

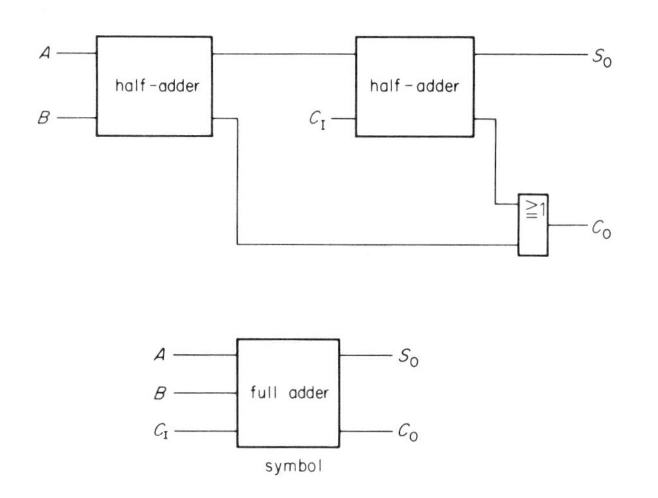


Figure 9.2 A full adder

To complete the addition process we need to be able to add the carry-in $C_1$ to the sum $S$ generated by the half-adder in figure 9.1. The complete addition will provide us with the *output sum* $S_O$ and, quite possibly, a further carry bit, $C_O$. The generation of the two carry bits was illustrated in the example in section 9.3. A *full-adder* circuit which combines the outputs of two half-adders is shown in figure 9.2, the *output carry* $C_O$ being obtained by OR-gating the carry outputs from the two half-adders.

## A Serial Adder

The numbers $A$ and $B$ may be part of a sequence of binary digits which are presented to the adder from, say, the store of a computer. The essential elements of the serial adder are shown in figure 9.3. In the case of a serial adder, the $2^0$ digits of both numbers are presented to inputs $A$ and $B$ of the adder, the sum appearing as output $S_O$. The carry-out signal resulting from this stage of the addition is stored in the *carry store* FFC.

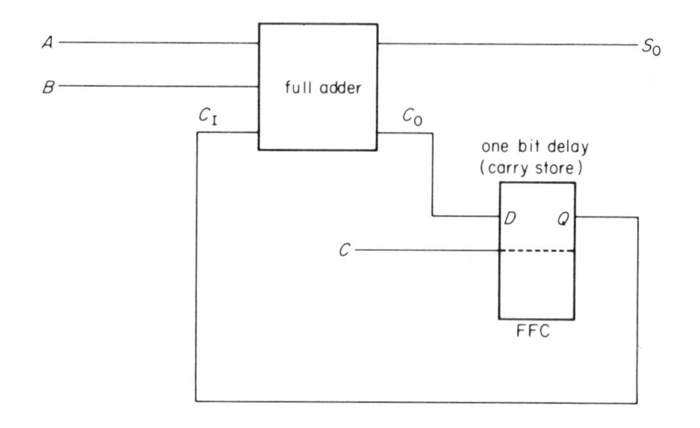


Figure 9.3 A serial adder

When the $2^1$ bits of numbers $A$ and $B$ are presented to the input of the adder, they are added together with the previous value of the carry-out signal generated by the $2^0$ addition. To delay the transfer of the $C_O$ bit by the correct time interval in order that its arrival at the $C_I$ input coincides with the $2^1$ bits of $A$ and $B$, the carry store is controlled by a clock signal. In figure 9.3 a D flip-flop is used as the carry store. More is said about the control of the flow of information in the following paragraph.

In serial machines the numbers $A$ and $B$ and the sum $S_O$ are stored in memory banks known as *shift registers* (see section 10.8). Each bit associated with numbers $A$ and $B$ is *shifted* into the adder under the control of a *clock pulse* or *shift pulse* C. By using a common shift pulse to shift data into inputs $A$ and $B$ as well as controlling the carry store, the correct delay is introduced to the carry bit so as to give the correct operation. If each number contains four bits, then four shift pulses are required to complete the addition.

### A Parallel Adder

In many circuits the time taken to carry out mathematical operations must be kept to a minimum. Parallel addition is used in these cases, since in parallel adders all the binary digits are simultaneously presented to the adder and the addition is carried out in one operation.

A block diagram of a 4-bit parallel full adder is shown in figure 9.4. A complete network of this kind in IC form is known as a *quad full adder*, and is contained in a 16-pin DIP. Inputs $A_1$, $A_2$, $A_3$, and $A_4$ correspond to the binary values of number $A$, $A_4$ being the *most significant bit* (m.s.b.). As can be seen from figures 9.3 and 9.4, parallel adders are more complex than serial adders. In the network shown, as each value of $A$ and $B$ is added, the carry is immediately applied to the next higher group of digits. Thus, the time taken to add two numbers together is only the time required for the carry signal to ripple through four stages. Even so, computer designers have produced devices known as *carry look-ahead adders* which are even faster in operation. These adders include additional logic circuitry to reduce the number of stages through which the carry must be propagated.

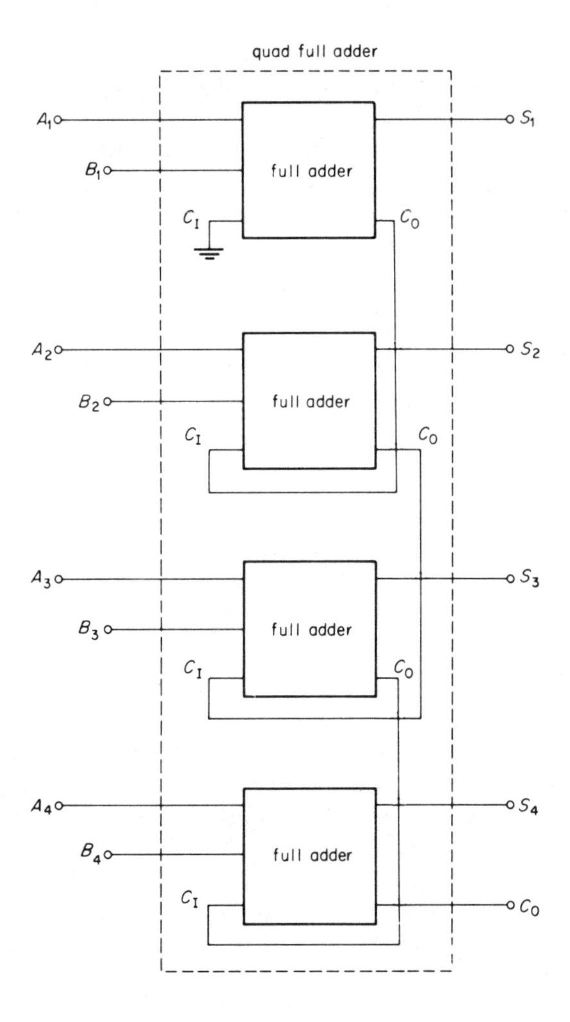


Figure 9.4 A parallel adder

## 9.5 BINARY SUBTRACTION

The process of subtraction is carried out by converting the number to be subtracted (the *subtrahend*) into a negative number, and the *difference* is formed by *adding* the negative subtrahend to the *minuend*. This process is illustrated in section 9.6.

## 9.6 NEGATIVE NUMBERS

A negative number is usually indicated by writing a 'minus' sign in front of the number. Unfortunately, electronic devices can only recognise the '0' and '1' signal levels, and cannot understand our minus sign. It is necessary, therefore, to develop techniques for the presentation of negative numbers in digital form.

The simplest method would be to use an additional bit in the code grouping, which we call the *sign bit*, which tells us whether the number is either positive or negative. We could, for example, allow the sign bit to assume the logical '0' value for positive numbers and allow it to be '1' for negative values. The remaining bits in the code group give the *magnitude* or *modulus* of the number. Such a number is said to be represented in *signed magnitude* (or *signed modulus*) form, several values being given below. In the examples given the sign bit is enclosed in brackets but, in a calculating machine, it would be identified simply by the fact that it is the m.s.b. in the number.

| decimal | binary |
|---|---|
| −02.5 | (1)0010.100 |
| +10.25 | (0)1010.010 |
| −00.125 | (1)0000.001 |

Note the apparently unusual practice of including the zeros above the most significant digit in some of the numbers. These zeros are known as *non-significant zeros*, since they do not contribute to the value of the number. This has been done deliberately because, in calculators and computers, the memory stores the value of each and every bit, even if it is zero. In order to store a number as small as decimal 0.125 we need three bits to the right of the decimal point, and to store decimal 10 we need four bits to the right. In addition, a sign bit is also required. Thus to store numbers in the range ±10.125 we need a binary storage capacity of eight bits (eight bits is frequently known as a *byte*).

The signed modulus notation is a convenient method of storing numbers, but it does not allow us to directly subtract two numbers from one another. A number can be subtracted from another by *adding* its *complement*, which is its equivalent negative value. To illustrate the complement notation we shall study how it operates in the *decimal* system. Let us first of all determine the complement of the decimal number 1 by subtracting it from zero. Here we use a five-digit code, the m.s.b. being the sign digit with '0' representing a positive number and '9' representing a negative number.

```
(0)0000
(0)0001
-------
                 subtract
(9)9999          signed complement of decimal − 1
-------
```

Hence we can differentiate between +9999 and −1 by the fact that, in the case of −1, the sign bit is (9). Hence, if the sign bit is (9), the number is stored in complement form.

Let us now look at the *binary* complement notation. Once again, to form the complement of the number, we subtract the positive value of the number from zero. This is illustrated below by forming the binary complement of 5 using a five-bit code, as follows

```
zero            (0)0000
decimal 5       (0)0101
                -------
                          subtract
                (1)1011   signed complement form of −5
                -------
```

The number so formed is known as the *true complement* or *two's complement* of the number. To verify that our result is correct, we shall subtract decimal 5 from decimal 9 in binary using two's complement addition.

```
decimal +9              (0)1001
decimal −5              (1)1011      signed 2's complement
                        -------      add
Overflow bit (lost)→ 1(0)0100        = difference
                        -------
```

The difference is (0)0100 or +4 (decimal). Note that in addition an overflow of '1' has occurred, but this is lost in the calculation since the storage capacity of our calculator is only five bits.

An alternative form of binary complement, known as the *one's complement* is also used. The 1's complement has a value which is less than the 2's complement by a factor of unity. The signed 1's complement of −5 is, therefore. (1)1010. Simple rules for the derivation of both the 1's and 2's complements are given below.

1's complement: Change all the 1s into 0s, and 0s into 1s.
2's complement: Commencing with the least significant bit (l.s.b.), leave all the digits up to and including the least significant '1' unchanged. All the more significant 0s are then changed into 1s, and 1s into 0s.
Alternatively, the 2's complement can be formed by adding '1' to the l.s.b. of the 1's complement.

Examples of the signed complement notation using a 7-bit data code are given below.

| decimal value | binary modulus | signed 2's complement | signed 1's complement |
|---|---|---|---|
| − 6 | (0)0110.000 | (1)1010.000 | (1)1001.111 |
| − 12.5 | (0)1100.100 | (1)0011.100 | (1)0011.011 |
| − 15.875 | (0)1111.111 | (1)0000.001 | (1)0000.000 |

## 9.7 BINARY MULTIPLICATION

Multiplication can be performed by a process of adding and shifting, as illustrated in the following problem in which the binary equivalents of the decimal numbers 5 and 6 are multiplied together.

```
  110     L (multiplicand) = 6
  101       M (multiplier) = 5
  ---
  110     partial product = L × 1
 000      partial product = L × 0
110       partial product = L × 1
-----
11110     product = sum of partial products
```

When multiplying numbers of different mathematical signs, the resulting sign bit can be generated by comparing the sign bits of the two numbers in a *not-equivalent* gate. If the sign bits are not equivalent to one another, the resulting sign bit will be a '1' (that is, the product has a negative sign). If they are equivalent, the output from the *not-equivalent* gate is '0'. Using this method, we must multiply the magnitudes of the two numbers and must not use complement notation.

**Table 9.4**

| *Decimal value* | *BCD codes* 8421 *BCD* (8) (4) (2) (1) | *excess-3 code* | 5421 *BCD* (5) (4) (2) (1) |
|---|---|---|---|
| 0 | 0 0 0 0 | 0 0 1 1 | 0 0 0 0 |
| 1 | 0 0 0 1 | 0 1 0 0 | 0 0 0 1 |
| 2 | 0 0 1 0 | 0 1 0 1 | 0 0 1 0 |
| 3 | 0 0 1 1 | 0 1 1 0 | 0 0 1 1 |
| 4 | 0 1 0 0 | 0 1 1 1 | 0 1 0 0 |
| 5 | 0 1 0 1 | 1 0 0 0 | 1 0 0 0 |
| 6 | 0 1 1 0 | 1 0 0 1 | 1 0 0 1 |
| 7 | 0 1 1 1 | 1 0 1 0 | 1 0 1 0 |
| 8 | 1 0 0 0 | 1 0 1 1 | 1 0 1 1 |
| 9 | 1 0 0 1 | 1 1 0 0 | 1 1 0 0 |
| unused combi-nations | 1 0 1 0 | 1 1 0 1 | 0 1 0 1 |
| | 1 0 1 1 | 1 1 1 0 | 0 1 1 0 |
| | 1 1 0 0 | 1 1 1 1 | 0 1 1 1 |
| | 1 1 0 1 | 0 0 0 0 | 1 1 0 1 |
| | 1 1 1 0 | 0 0 0 1 | 1 1 1 0 |
| | 1 1 1 1 | 0 0 1 0 | 1 1 1 1 |

## 9.8 BINARY DIVISION

As with decimal division, binary division can be performed by a process of subtracting and shifting.

## 9.9 BINARY–DECIMAL CODES

Where a man-to-digital system relationship exists, it is vital to establish codes which can be understood by both. The digital system can handle pure binary numbers with ease, but man experiences some difficulty in dealing with binary numbers. In an attempt to minimise the problems of communication involved, several codes known as *binary-coded decimal* (*BCD*) *codes* have been developed. Using these codes, the presentation of decimal information in binary form is relatively simple. Three such codes are listed in table 9.4.

Each code uses ten of the sixteen possible combinations, the remaining six combinations being unused. The 8421 BCD code binary digits have weights of 8, 4, 2 and 1, respectively, and use the first ten groups of the natural binary code sequence listed in table 9.1. When reference is made to *the* BCD code, the 8421 BCD code is referred to.

The excess-3 code is generated by adding the binary equivalent of decimal 3 to each of the groups in the 8421 BCD code. An advantage of the excess-3 code over the 8421 BCD code is the ease with which some mathematical calculations can be carried out.

Yet another code, the 5421 BCD code, is shown in table 9.3. In this code the weight of the m.s.b. is five, so that the decimal number eight is made up of the group

$$(1 \times 5) + (0 \times 4) + (1 \times 2) + (1 \times 1)$$

Many other forms of BCD code exist, some having negative weights, the 642(–3) BCD code being an example. In that code the decimal number five is represented by the binary group 1011 and decimal seven by 1101.

For values greater than decimal 9, the BCD values or *weights* change by a factor equivalent to decimal ten. Thus the first four bits to the left of the binary point in the 8421 BCD code have weights of 8, 4, 2 and 1 respectively; the next more significant group of four bits have weights of 80, 40, 20 and 10 respectively and so on. As an example, the decimal number 872.5 is represented in the 8421 BCD code as 1000 0111 0010.0101.

In addition to positively weighted codes (the 8421 BCD code being an example) and negatively weighted codes (the 642(–3) BCD code for example), there are others including unweighted codes. Unweighted codes are those in which a decimal value cannot be allocated to each bit. Examples of these codes are given in table 9.5.

Both codes in table 9.5 are examples of *unit-distance codes*. The *code distance* is the number of bits which change between any

**Table 9.5**

| *Decimal value* | *Gray code* (repeat: 15 → 0) | *Walking code or creeping code* (repeat: 9 → 0) |
|---|---|---|
| 0 | 0000 | 00000 |
| 1 | 0001 | 00001 |
| 2 | 0011 | 00011 |
| 3 | 0010 | 00111 |
| 4 | 0110 | 01111 |
| 5 | 0111 | 11111 |
| 6 | 0101 | 11110 |
| 7 | 0100 | 11100 |
| 8 | 1100 | 11000 |
| 9 | 1101 | 10000 |
| 10 | 1111 | |
| 11 | 1110 | |
| 12 | 1010 | |
| 13 | 1011 | |
| 14 | 1001 | |
| 15 | 1000 | |

adjacent pair of numbers. In the case of both codes in table 9.5, only one bit changes between any pair of adjacent values. For example, in the Gray code, when the value increases from 6 to 7, the only change is in the least significant bit; when the value changes from 6 to 5 the only change in the Gray code value is in the second bit. Also when the value changes from 15 to zero (on the completion of the sequence) the only bit to change is the most significant one.

The *walking code* or *creeping code* gives the appearance of 1s creeping from right to left, to be finally replaced by 0s 'walking' through. This type of code is also known as a *Johnson* code (see also section 10.11).

Special codes are used for transmitting *alpha*betical and *numeri*cal data (alphanumeric codes), a simple six bit code of this type being given in table 9.6. In this code the two left-hand bits are used to distinguish between numerical and alphabetical information, while the four right-hand bits are coded in the 8421 BCD code. The code given is limited in usefulness since it deals only with capital letters and does not deal with punctuation marks or with typewriter instructions (such as carriage return and line feed).

**Table 9.6**

| *Decimal value* | *Code* | *Letter* | *Code* | *Letter* | *Code* | *Letter* | *Code* |
|---|---|---|---|---|---|---|---|
| 0 | 00 0000 | A | 01 0000 | K | 10 0000 | U | 11 0000 |
| 1 | 00 0001 | B | 01 0001 | L | 10 0001 | V | 11 0001 |
| 2 | 00 0010 | C | 01 0010 | M | 10 0010 | W | 11 0010 |
| 3 | 00 0011 | D | 01 0011 | N | 10 0011 | X | 11 0011 |
| 4 | 00 0100 | E | 01 0100 | O | 10 0100 | Y | 11 0100 |
| 5 | 00 0101 | F | 01 0101 | P | 10 0101 | Z | 11 0101 |
| 6 | 00 0110 | G | 01 0110 | Q | 10 0110 | | |
| 7 | 00 0111 | H | 01 0111 | R | 10 0111 | | |
| 8 | 00 1000 | I | 01 1000 | S | 10 1000 | | |
| 9 | 00 1001 | J | 01 1001 | T | 10 1001 | | |

In general, there is no 'best' code for all applications. The code selected is usually a compromise between many factors including transmission efficiency, accuracy and circuit simplicity.

## 9.10 ERROR DETECTION

When a binary code group is being transmitted from one point to another within a system, it is possible that one or more bits may either be lost ('drop-outs') or picked up ('drop-ins'). When this occurs, the code group contains an error. Several codes have been developed which not only detect the errors but, in some cases, allow us to correct the errors. In the simplest form of error detection we introduce an additional bit to the code group, known as the *parity bit*, which is *redundant* in terms of information transmission.

In an *odd-parity check* system, the parity bit makes the total number of 1s in the code group an odd sum and in the case of an *even-parity check* it makes the sum an even number. Parity checks can be used with any type of code, and the 8421 BCD code together with the parity bit $P$ for both odd and even parity are listed in table 9.7. The type of parity check selected depends on the application, since each has its advantages; for example, a feature of an odd-parity check is that no number is represented by a series of zeros.

**Table 9.7**

| *Decimal value* | *Odd parity* 8421$P$ | *Even parity* 8421$P$ |
|---|---|---|
| 0 | 00001 | 00000 |
| 1 | 00010 | 00011 |
| 2 | 00100 | 00101 |
| 3 | 00111 | 00110 |
| 4 | 01000 | 01001 |
| 5 | 01011 | 01010 |
| 6 | 01101 | 01100 |
| 7 | 01110 | 01111 |
| 8 | 10000 | 10001 |
| 9 | 10011 | 10010 |

A parity-bit generator for a code group which is presented in parallel mode, that is, all the bits are presented simultaneously, is shown in figure 9.5a. The output from the NOT-EQUIVALENT gate G3 is the even-parity bit associated with the four inputs. From table 9.7 we see that the odd and even-parity bits are the logical complements of one another, so that the output from G4 in the figure is the odd-parity bit associated with the four inputs.

A simple parity-check circuit is shown in figure 9.5b. Here, $P_1$ is the parity bit associated with the incoming code group and $P_2$ is a parity-check bit generated at the *receiving point* by a circuit of the type in figure 9.5a. If the incoming parity bit is correct, then the output from the parity checker is zero.

A parity generator for a serial binary number could, quite simply, be a trigger flip–flop whose output is initially set to zero. The even-parity bit is obtained from the $Q$-output of the flip–flop, and the odd-parity bit from the $\overline{Q}$-output.

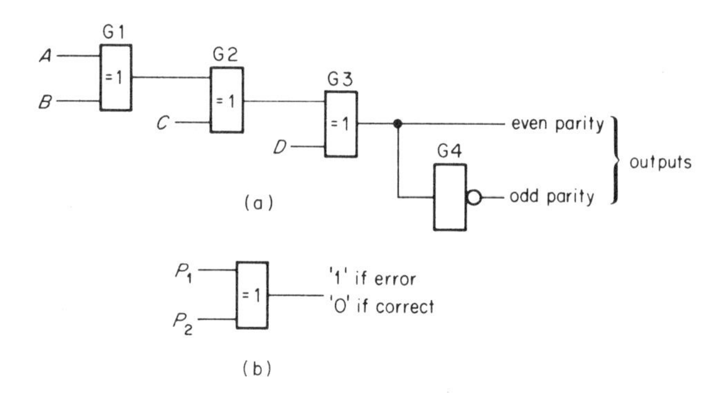


Figure 9.5 Parity-bit generators

## BCD Illegal Code Checker

We saw in table 9.4 that out of the sixteen possible combinations of four bits, we needed only ten in the BCD code. The remainder are unused and are illegal code groups. Let us consider the design of a network which gives an indication of the existence of any one of the unused or illegal code groups in the 8421 BCD code.

From table 9.4 we note that the following groups should not exist: 1010, 1011, 1100, 1101, 1110 and 1111, the groups representing the sequence $ABCD$ where $A$ is the m.s.b. Mapping these combinations on the Karnaugh map in figure 9.6a, we see that the logical expression representing them is

$$\text{error} = A.B + A.C = A.(B + C)$$

A block diagram of a network which generates this function is shown in figure 9.6b, the circuit giving a '1' output when an illegal code group is presented to its input. Note that signal $D$ in the BCD code is not required in this circuit.

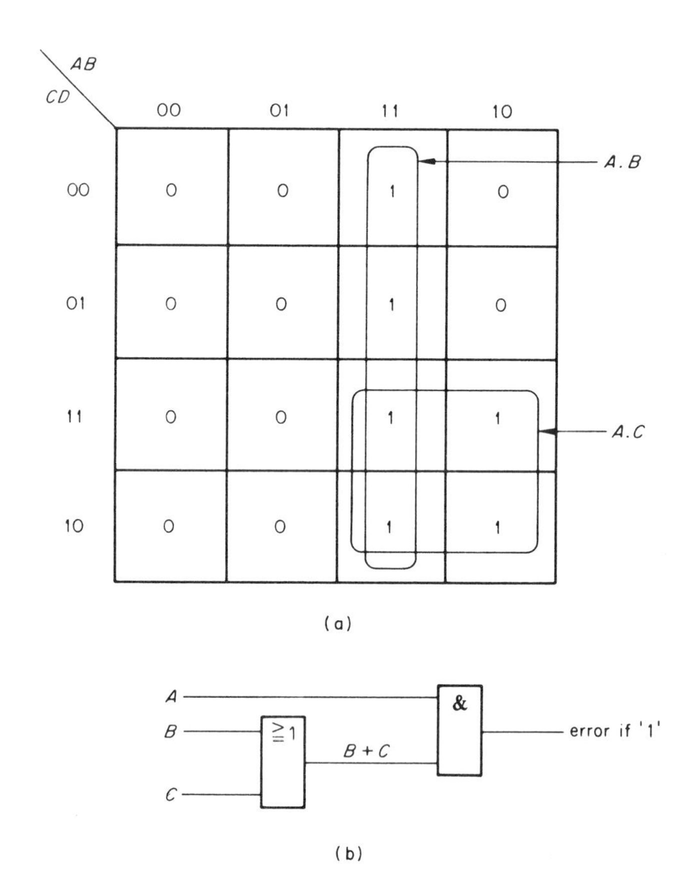


Figure 9.6 An illegal code-checking circuit for the 8421 BCD code

## PROBLEMS

**9.1** Convert the decimal number 92.875 into its pure binary equivalent.

**9.2** Convert the pure binary number 11011.111 into its decimal equivalent.

**9.3** Convert the hexadecimal number AC8 into (a) its binary equivalent, (b) its octal equivalent and (c) its decimal equivalent.

**9.4** Convert the decimal numbers 26.625 and 15.5 into their pure binary equivalent and add the numbers together in binary.

**9.5** Convert the following numbers into their 2's complement form: −12.25, −3.75, 387, −100.

**9.6** Perform the following subtraction in pure binary using the method of complement addition: 29.75 − 12.625.

**9.7** Express the following decimal numbers in the 8421 BCD code: 37.5, 897, 2087.5.

**9.8** Discuss the features of unit distance codes for use in (a) position-sensing systems, (b) binary arithmetic systems.

**9.9** The three signals shown in figure 9.7 are applied to a parity-bit generator. If odd parity is used, draw the output waveform from the generator.

**9.10** Design a logic network using only NAND gates which checks for illegal code groups in the 5421 BCD code in table 9.4.

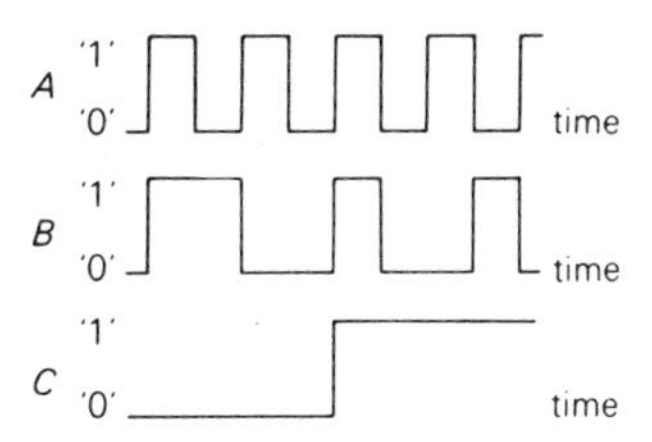


Figure 9.7

# 10 Counter and Shift Register Circuits

## 10.1 A PURE BINARY ASYNCHRONOUS COUNTER

When a group of flip–flops are connected together so that they store related information, they are known collectively as a *register*. Certain types of register can be used for the purpose of counting pulses and are known as *counters*. An *asynchronous counter* or *serial counter* is one in which the pulses are applied at one end of the counter and the process of adding each pulse has to be completed before the 'carry bit' is propagated to the following stage. The following stage has then to add the carry bit to the number in that stage. That is, the carry bit appears to 'ripple' through the length of the counter until the count is complete. As a result, asynchronous counters are sometimes known as *ripple-through counters*.

A three-stage asynchronous counter using J–K flip–flops connected as T flip–flops is illustrated in figure 10.1a. All the stages of the counter are initially set to zero by the application of a signal to the reset line during the period of time when the input signal is zero (see section 8.6). As explained in section 8.6, the leading edge of the input pulse (that is, when the input signal changes from '0' to '1') has no effect on the state of FFC, so that all outputs remain at '0'. When the input pulse falls to zero, it causes the output of FFC to change to '1'. The change in signal at the output of FFC is applied to the 'clock' input of FFB but, as the change is a '0' to '1' change it has no effect on the operating state of FFB, that is, its output remains as '0'. The code sequence generated by the counter is listed in table 10.1, and the change described above corresponds to the change from the first row of the table to the second row.

At the end of the second pulse the output of FFC once more changes, this time from '1' to '0'. Since this corresponds to the trailing edge of the pulse applied to the clock input line of FFB, the output of FFB changes from '0' to '1'. This change in FFB output has no effect on the state of FFA, so that the outputs now are $A = 0$, $B = 1$, $C = 0$. Expanding the time scale associated with this transition as shown in the inset to figure 10.1b, we see that it takes a finite time $t_x$ for the transition to ripple through from FFC to FFB. This period of time is very small, but finite.

The count proceeds in the manner prescribed in table 9.1, the changes in the outputs occurring when the input to individual

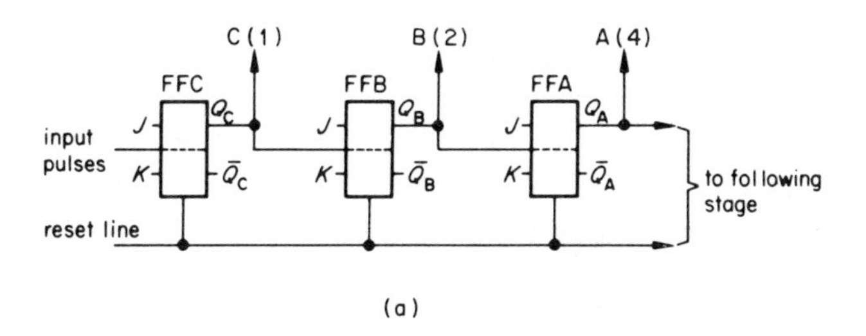


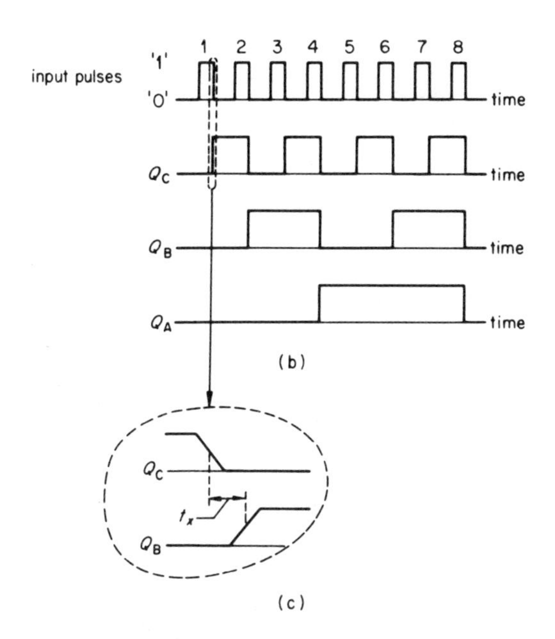


Figure 10.1 A ripple-through pure binary counter

flip–flops changes from '1' to '0'. After seven pulses have been applied to the counter, all the outputs are at the '1' level, so that the eighth pulse causes them all to fall to zero. As shown in figure 10.1c, the time taken for each change to propagate to the following stage is $t_x$, so that it takes $3t_x$ seconds before all the outputs from the counter have settled down to their steady-state values. As a result of this mode of operation, the maximum counting rate is restricted by the time delays in the counter and for a counter with $n$ stages the delay time before the final state of the counter can be 'read' after the application of a pulse is $nt_x$ seconds. This time delay is very much reduced in the synchronous counters described in section 10.4.

## 10.2 A BIDIRECTIONAL PURE BINARY ASYNCHRONOUS COUNTER

The type of counter described above is referred to as a *forward counter* or as an *'up' counter*, since it counts 'up' from zero. It is sometimes convenient to count from some predetermined value down to zero. A counter operating in this mode is known as a *reverse counter* or as a *'down' counter*.

The design philosophy of a 'down' counter can be deduced from table 10.1. Here we see that the *total* value stored in both the $Q$ and $\overline{Q}$ outputs is constant and is equal to decimal 7 (this takes the weights of each bit into account). Thus, the initial value stored in the $Q$ outputs is zero and that stored in the $\overline{Q}$ outputs is 7. After the first pulse, the $Q$ outputs store 1 and the $\overline{Q}$ outputs store 6, and so on. Clearly it is possible to cause our counter to count 'down' either if we monitor the $\overline{Q}$ outputs rather than the $Q$ outputs or if we use the $\overline{Q}$ output of an earlier stage to trigger the following stage. By

**Table 10.1**

| *Pulse* | *A* (4) | *B* (2) | *C* (1) | | $\overline{A}$ (4) | $\overline{B}$ (2) | $\overline{C}$ (1) | |
|---|---|---|---|---|---|---|---|---|
| initial condition | 0 | 0 | 0 | repeat | 1 | 1 | 1 | repeat |
| 1 | 0 | 0 | 1 | | 1 | 1 | 0 | |
| 2 | 0 | 1 | 0 | | 1 | 0 | 1 | |
| 3 | 0 | 1 | 1 | | 1 | 0 | 0 | |
| 4 | 1 | 0 | 0 | | 0 | 1 | 1 | |
| 5 | 1 | 0 | 1 | | 0 | 1 | 0 | |
| 6 | 1 | 1 | 0 | | 0 | 0 | 1 | |
| 7 | 1 | 1 | 1 | | 0 | 0 | 0 | |
| 8 | 0 | 0 | 0 | | 1 | 1 | 1 | |

the use of a suitable electronic gating system, we can design a *reversible counter* or *bidirectional counter* which can count either up or down.

A pure binary asynchronous reversible counter is shown in figure 10.2, in which the signal applied to line $U$ controls whether the counter operates in the 'up' mode or the 'down' mode. If $U = 1$, the upper AND gates (that is, G1, G2, G3) are activated. Since one input of each of the upper AND gates is connected to a $Q$ output line, the counter counts 'up' when $U = 1$ since the changes in the $Q$ outputs are gated forward. When $U = 0$, the lower AND gates (that is, G4, G5, G6) are activated. Since one input of each of these gates is connected to a $\bar{Q}$ output, the counter counts 'down' when $U = 0$.

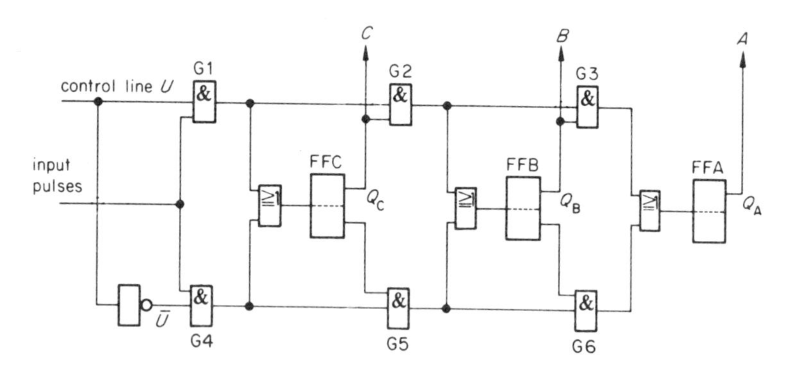


Figure 10.2 A reversible pure binary counter

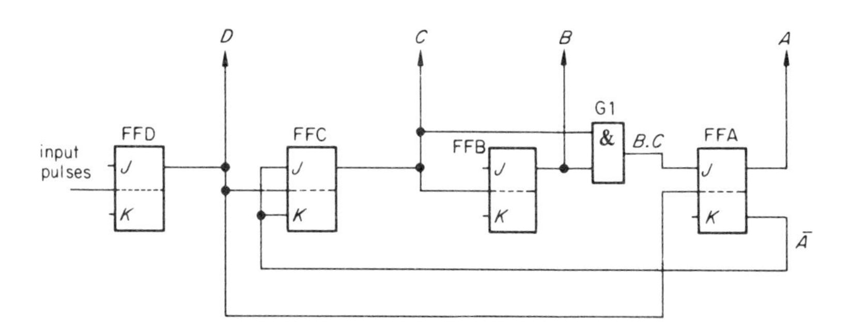


Figure 10.3 A ripple-through 8421 BCD counter

## 10.3 AN 8421 BCD ASYNCHRONOUS COUNTER

A popular form of 8421 BCD 'up' counter using J–K flip–flops is shown in figure 10.3 and the code sequence generated is listed in table 10.2. Note that FFB and FFD are connected to operate as conventional T flip–flops.

**Table 10.2**

| *Pulse* | *ABCD* |
|---|---|
| initial value | 0000 |
| 1 | 0001 |
| 2 | 0010 |
| 3 | 0011 |
| 4 | 0100 |
| 5 | 0101 |
| 6 | 0110 |
| 7 | 0111 |
| 8 | 1000 |
| 9 | 1001 |
| 10 | 0000 |

(repeat)

Initially, with all outputs reset to zero, the signal at the $\bar{A}$ terminal is '1'. This is fed back to the $J$ and $K$ inputs of FFC so that, initially, it operates in a T flip–flop mode. With the circuit so energised, the signal $B.C$ from G1 is zero, making FFA inoperative. Thus, FFD, FFC and FFB operate as conventional trigger flip–flops, which then count 'up' in the normal pure binary sequence for the first seven pulses (see table 10.2). During this counting period (in fact after the sixth pulse), output $B.C$ becomes '1' and prepares FFA for operation. After the seventh pulse, the output from FFD once more becomes '1'. Since signal $D$ is the clock source for FFA, it allows the '1' at the output of G1 to be entered into the 'J' master stage of FFA. At the end of the eighth pulse outputs $B$, $C$ and $D$ fall to zero. Since the change in the latter signal corresponds to the 'trailing' edge of FFA clock pulse, the '1' stored in the master stage of FFA is gated to output $A$. Thus after eight clock pulses $A = 1, B = 0, C = 0, D = 0$. At the same instant of time output $\overline{A}$ falls to zero, thereby inhibiting the operation of FFC and FFB. The ninth pulse causes output $D$ to become '1' once more. The next pulse, the tenth pulse, results in $D$ becoming zero again and this change is applied to the clock input of FFA. Since $B = C = 0$ at this time, the signal applied to the $J$-input of FFA is '0', so that the tenth input pulse causes a '0' ($= B\,.\,C$) to appear at output $A$. Thus, once more all the outputs are again zero, and the cycle is recommenced by the eleventh pulse.

If FFA in figure 10.3 is replaced by a flip–flop with two $J$-inputs (see also figure 8.6), the necessity for gate G1 is removed, as the AND function is carried out inside the flip–flop.

## 10.4 A REASON FOR SYNCHRONOUS COUNTERS

In synchronous counters, the counting sequence is controlled by means of a clock pulse and all the changes of the outputs of *all* flip–flops occur in synchronism. This effectively eliminates the large propagation delay associated with ripple-through counters, as mentioned in section 10.1. Master–slave flip–flops are invariably used in synchronous counters to avoid the possibility of oscillation and instability when feedback connections are made in the completed counter. In this mode of operation, the appropriate input signals are simultaneously gated into the master stages of all the flip–flops in the counter. When the input pulse falls to the '0' level, the new values of the count are transmitted synchronously to the outputs of the flip–flops.

## 10.5 A SYNCHRONOUS PURE BINARY COUNTER

One form of synchronous pure binary counter is shown in figure 10.4, table 10.3 giving the sequence of events taking place in the counter, and from which we can deduce the design principles of the counter.

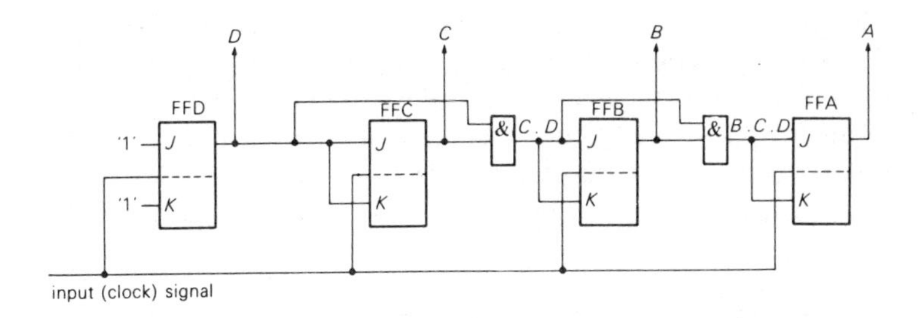


Figure 10.4 A synchronous pure binary counter

**Table 10.3**

| Pulse | A (8) | B (4) | C (2) | D (1) |
|---|---|---|---|---|
| Initial condition | 0 | 0 | 0 | 0 |
| 1 | 0 | 0 | 0 | 1 |
| 2 | 0 | 0 | 1 | 0 |
| 3 | 0 | 0 | 1 | 1 |
| 4 | 0 | 1 | 0 | 0 |
| 5 | 0 | 1 | 0 | 1 |
| 6 | 0 | 1 | 1 | 0 |
| 7 | 0 | 1 | 1 | 1 |
| 8 | 1 | 0 | 0 | 0 |
| 9 | 1 | 0 | 0 | 1 |
| 10 | 1 | 0 | 1 | 0 |
| 11 | 1 | 0 | 1 | 1 |
| 12 | 1 | 1 | 0 | 0 |
| 13 | 1 | 1 | 0 | 1 |
| 14 | 1 | 1 | 1 | 0 |
| 15 | 1 | 1 | 1 | 1 |
| 16 | 0 | 0 | 0 | 0 |

(repeat)

We see from the table that the output of FFD must change after every input pulse. This calls for FFD to be connected to operate as a T flip–flop. The output of FFC is seen to change *following* the condition that $D = 1$ (that is, following odd-numbered pulses). This change is brought about by driving both the $J$-input and the $K$-input of FFC by signal $D$. Also, the output of FFB must change state *following* the condition that the logical combination $C \cdot D = 1$ is satisfied. This occurs after pulses 3, 7, 11 and 15 have been received. Similarly, the output of FFA must change *after* the condition $B \cdot C \cdot D = 1$ has been satisfied, that is, after pulses 7 and 15 have been received.

If J–K flip–flops with multiple $J$ and $K$-inputs are used, then the AND gates associated with FFA and FFB are not required, as this function can be generated internally (see figure 8.6).

## 10.6 A REVERSIBLE SYNCHRONOUS PURE BINARY COUNTER

One form of reversible counter is shown in figure 10.5, the 'up' or 'down' mode being selected by the signal applied to line $U$ which is the up/down control line. If $U = 1$, the counter operates in a count-up mode and when $U = 0$ it counts down.

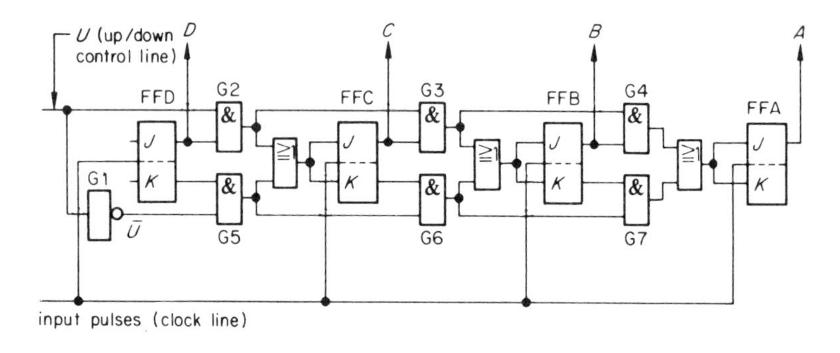


Figure 10.5 A reversible synchronous pure binary counter

When $U = 1$, the '0' output from gate G1 inhibits the operation of gates G5, G6 and G7; the '1' signal on the $U$ line permits G2, G3 and G4 to function. In this condition, the part of the circuit that is operational is generally similar to the counter in figure 10.4. Consequently the circuit operates in a count-up mode. When $U = 0$, gates G2 to G4 are inhibited and gates G5 to G7 are opened to the flow of information. In this operating state, the data stored in the $\overline{Q}$ outputs is transmitted forward and the counter counts down.

## 10.7 A SYNCHRONOUS 8421 BCD COUNTER

The general arrangement of the synchronous counter in figure 10.6 is similar to the asynchronous BCD counter described earlier in section 10.3. The operation of the circuit is as follows.

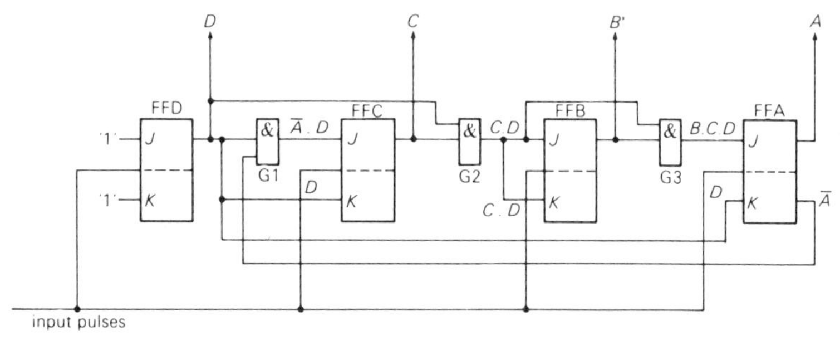


Figure 10.6 A synchronous 8421 BCD counter

Assuming that the outputs are all initially zero, the '1' signal at output $\overline{A}$ is fed back to G1 to allow it to pass signals. The circuit is then electrically equivalent to the synchronous counter in figure 10.4. The network counts up synchronously for the first eight pulses but, at the end of the eighth pulse, output $A$ becomes '1' and $\overline{A}$ falls to '0'. This action inhibits the operation of G1 and prevents further signals from being propagated to the $J$-input of FFC. The ninth pulse causes the output of FFD to change to '1', when the conditions in the counter are $A = D = 1$, $B = C = 0$. At the end of the tenth pulse the previous value of $D$, '1', is gated into the $K$-input of FFA to cause output $A$ to fall to zero. At the same time, the tenth pulse also causes output $D$ to fall to zero, and the counter is reset to its initial condition of $A = B = C = D = 0$.

## 10.8 SHIFT REGISTERS AND RING COUNTERS

A register is simply an array of flip–flops used for the storage of binary data and a *shift register* is one which is designed so that the data may be 'shifted' along the register in either direction, that is, either to the right or to the left. A *ring counter* is a shift register which is connected in the form of a continuous ring, the input to the 'beginning' of the ring being a logical function of the signals at one or more points in the register.

## 10.9 A SERIAL-INPUT, SERIAL-OUTPUT SHIFT REGISTER

Two basic circuits for shift registers are shown in figure 10.7. In the circuits shown, the data to be stored in the register is shifted into the register in a serial fashion from the left-hand end, and is shifted out at the right-hand end. Let us consider the operation of figure 10.7a.

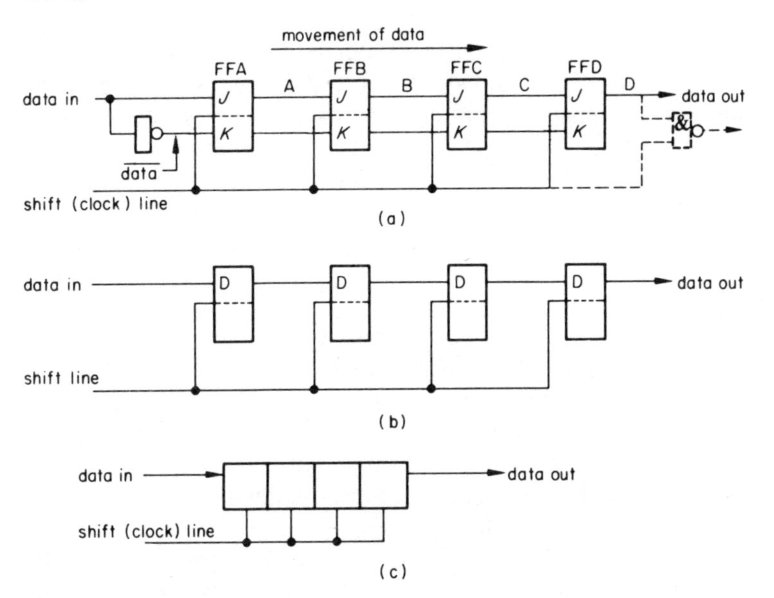


Figure 10.7 Serial-input, serial output shift registers

Assume that outputs *A*, *B*, *C* and *D* are initially zero. The data presented at the input line is logically inverted by the NOT gate, so that complementary signals are presented to the *J* and *K* lines of the first flip–flop. Suppose now that a '1' signal is applied to the DATA IN line and that a pulse is applied to the clock (SHIFT) line of the register. At the end of the clock pulse the '1' is transferred to output *A*. Meanwhile, outputs *B*, *C* and *D* remain at '0', since the input signals applied to those flip–flops were '0' during the period that the clock pulse was at the '1' level. The result of the above operation is shown in the first two rows of table 10.4. The '1' injected above is marked with an asterisk in table 10.4 to indicate its movement through the register.

**Table 10.4**

| *Clock pulse* | *Signal on data line* | *State of outputs* *A* | *B* | *C* | *D* | |
|---|---|---|---|---|---|---|
| initial condition | | 0 | 0 | 0 | 0 | |
| 1 | 1* | 1* | 0 | 0 | 0 | one complete shift cycle (pulses 1–4) |
| 2 | 0 | 0 | 1* | 0 | 0 | |
| 3 | 0 | 0 | 0 | 1* | 0 | |
| 4 | 1 | 1 | 0 | 0 | 1* | |
| 5 | 1 | 1 | 1 | 0 | 1 | |
| 6 | 1 | 1 | 1 | 1 | 0 | |

Thus we have shifted a '1' from the data line into FFA. Also we have shifted the '0' formerly stored in FFA into FFB, as well as shifting the '0' in FFB into FFC and the '0' in FFC into FFD. Hence *all* the information stored in the register has been transferred one step to the right.

If, now, the signal on the data line is reduced to zero and another shift pulse is applied, then this '0' is shifted into FFA at the end of the pulse. This change is brought about by shifting a '1' into the $\overline{A}$ output and at the same time the 1* is shifted into FFB.

Since the shift register has four stages, it can store a binary *word* of four bits, and requires four clock pulses to shift a new word into the register. Hence, after four clock pulses we have serially shifted out of FFD the word 0000 and then serially shifted in the word 1001. In the register in figure 10.7a, we say that the information has been *shifted up* or *shifted to the right*.

We can continue to shift data into and out of the register and table 10.4 is continued beyond one cycle to show the effect of shifting the first two 1s of the next word into the store. This has the effect of shifting the two right-hand bits stored at the end of the first cycle of the register.

In the form of circuit shown in figure 10.7a, the output from FFD will be a continuous signal, either a '1' or a '0', for the whole of the clock cycle. If we need the output in the form of a pulse train, then it can be obtained by AND-gating the output of FFD with the clock pulse as shown in broken lines in figure 10.7a. The interconnections between the flip–flops can be simplified by the use of D-type elements, as shown in figure 10.7b. These flip–flops have internal NOT gates which generate the $\overline{\text{DATA}}$ information internally.

A symbol used to represent serial registers is shown in figure 10.7c.

Data can be shifted to the *left* if, in figure 10.7a, the output from FFD is used as the input to FFC, the output from FFC is used as the input to FFB, and so on. The 'input' signal would then be fed into the $J$ and $K$-terminals of FFD and the 'output' would be taken from the $Q$-terminal of FFA. You would find it interesting to design a shift register which shifts left, or one which shifts either to the left or to the right, the direction of data movement being controlled by a logic signal (say '1' for shift right and '0' for shift left).

In the circuit in figure 10.7a, the data from each output can be extracted in parallel, that is, simultaneously by connecting outputs $A$, $B$, $C$ and $D$ to separate indicator lamps. Using additional logic circuitry, data can be input in parallel to the register. You would find it useful to design a logic network which fulfills the latter function.

## 10.10 DYNAMIC SHIFT REGISTERS

A unique property of the MOSFET is the facility that it can be used as a dynamic memory element, as was described in section 8.10. When used in conjunction with other MOSFETs, simple and reliable shift registers can be constructed in monolithic IC form, the basis of one type being shown in figure 10.8.

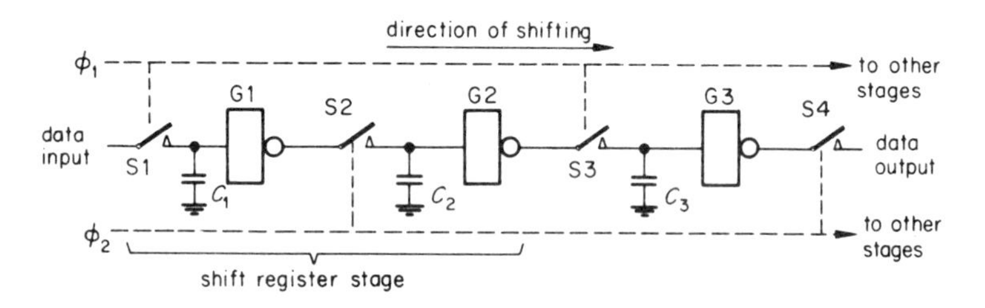


Figure 10.8 The basis of a MOS dynamic shift register

The memory elements are the invertor stages G1, G2, G3, etc., together with the parasitic gate capacitors $C1$, $C2$, $C3$, etc. A complete shift register stage comprises two of these memory elements, together with two MOST switches which are turned *on* and *off* by signals $\phi_1$ and $\phi_2$. The control lines $\phi_1$ and $\phi_2$ are energised by a *two-phase pulse supply*, the timing of $\phi_1$ and $\phi_2$ being such that the two switches associated with each memory are never closed simultaneously.

When $\phi_1$ is energised by a '1' signal, switch S1 is closed and the data on the input line is transferred to capacitor $C1$. Let us suppose that the data signal is logic '1'. This signal charges $C1$, and the inverting action of G1 causes the output of that gate to be '0'. At the end of the $\phi_1$ pulse, S1 opens and a logic '1' signal is applied to line $\phi_2$. This causes S2 to close, and the '0' signal at the output of G1 discharges $C2$. The inverting action of G2 causes the output of that gate to be logic '1'. Hence, in a complete cycle of events $\phi_1$ and $\phi_2$ become logic '1' in turn and cause the input data to be transferred into the first stage of the register.

During the period of time that input data is transferred into G1, the data stored in G2 is transferred into G3. Clearly, signals $\phi_1$ and $\phi_2$ cause all the data stored in the register to be shifted to the right.

A diagram of one section of one form of dynamic shift register is shown in figure 10.9. In this circuit TR1 replaces S1 in figure 10.8, TR4 replaces S2, and TR7 replaces S3. Gate G1 in figure 10.8 is replaced in figure 10.9 by TR2 and TR3, and G2 is replaced by TR5 and TR6.

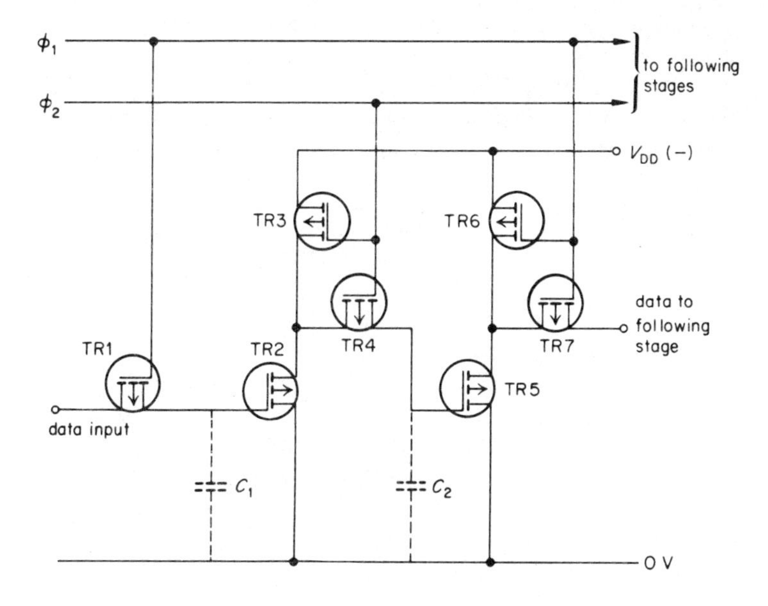


Figure 10.9 One stage of a dynamic shift register

## 10.11 RING COUNTERS

A ring counter is simply a shift register whose input is derived directly from its output, in the manner shown in figure 10.10.

Let us assume that stage A of figure 10.10a initially stores a '1', and that all other outputs are zero. The first shift pulse applied to the counter causes the '1' to move from stage A to stage B, and the 0s in stages B and C to move into stages C and D, respectively. Due to the feedback connection, the '0' in D is fed back into stage A. In this way the single '1' circulates continuously around the register in the manner shown in figure 10.10b.

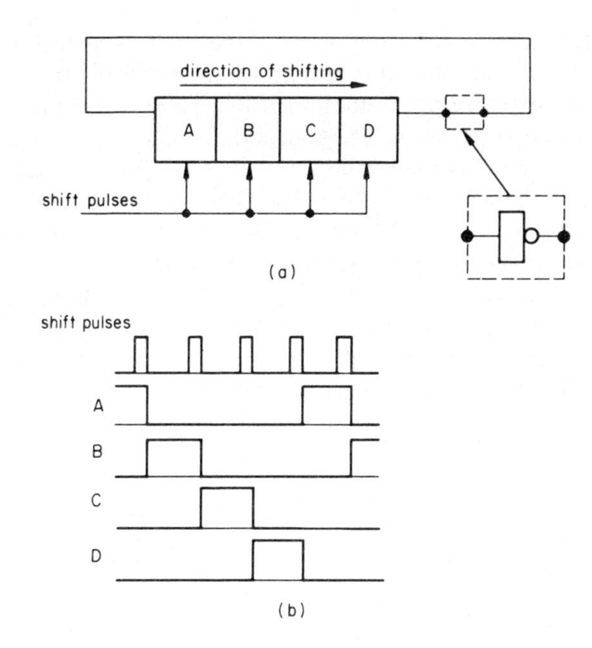


Figure 10.10 A ring counter

A feature of the ring counter is that the value of the number stored is easily decoded, since when a '1' appears at output A it is equivalent to a count of zero, when it appears at B it is equivalent to a count of unity, when at C it is equivalent to a count of two, and when at D three pulses have been applied. This means that the *cycle length* of the code generated is four for a four-stage shift register. A ring counter which operates in a decimal code would need ten flip–flops.

As we have seen in the chapter on counters, it is possible to generate a cycle length of $2^4 = 16$ with four binary stages. Evidently, although the ring counter provides us with a simple method of 'reading' the number stored, it is uneconomic in its use of flip–flops.

### Starting Conditions in a Ring Counter

When the ring counter is first switched on, there is no guarantee that stage A will contain a '1' and that the other stages will contain zeros. A method sometimes adopted in circuits using this type of counter is to correct the situation at the end of the first complete cycle of operations by including an additional gate in the feedback network. The operation will be correct after the first cycle. The operation of such a circuit is outlined below.

We have seen that a '1' must be fed into stage A *following* the condition that the outputs of stages A, B, C and D are *all* zero. If we use the feedback arrangement in figure 10.11a in which the input of stage A is energised by a NOR gate whose inputs are $A$, $B$, $C$ and $D$, then a '1' will be fed into A following the condition that $A = B = C = D = 0$. Alternatively we can use the feedback arrangement in figure 10.11b.

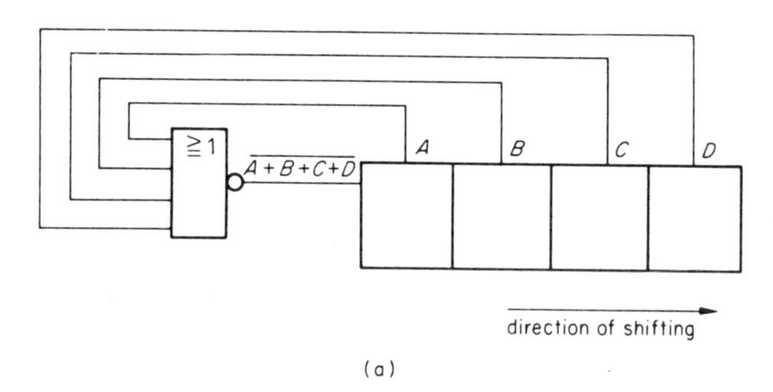


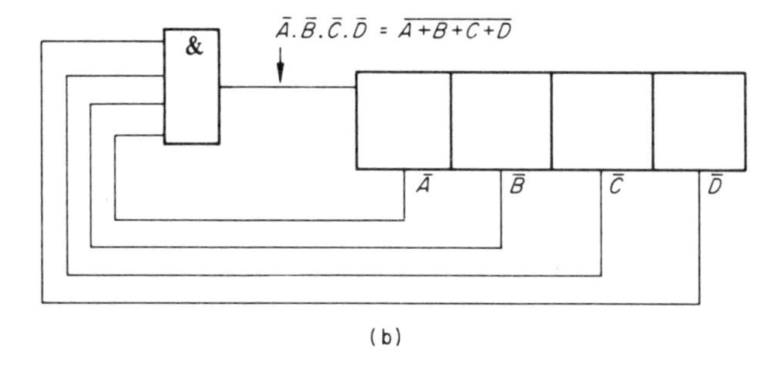


Figure 10.11 Methods of ensuring the correct conditions in a ring counter

## A Twisted Ring Counter

By including the NOT gate shown in the inset to figure 10.10a in the feedback loop of the counter, the input to stage A is the logical function $\bar{D}$. This has the effect of increasing the cycle length to eight, that is, the code cycle length is double that of an otherwise equivalent ring counter. Offset against this is the fact that it is slightly more difficult to decode into decimal the binary value stored in the counter. The code sequence generated by the counter, known as a *twisted ring counter* due to the inversion in the feedback loop, commencing with the group 0000, is given in table 10.5.

**Table 10.5**

| *Decimal value* | *A* | *B* | *C* | *D* |
|---|---|---|---|---|
| 0 | 0 | 0 | 0 | 0 |
| 1 | 1 | 0 | 0 | 0 |
| 2 | 1 | 1 | 0 | 0 |
| 3 | 1 | 1 | 1 | 0 |
| 4 | 1 | 1 | 1 | 1 |
| 5 | 0 | 1 | 1 | 1 |
| 6 | 0 | 0 | 1 | 1 |
| 7 | 0 | 0 | 0 | 1 |
| | 0 | 0 | 0 | 0 |

In this case the logical complement of $D$ is fed into A, so that $A$ becomes '1' following the condition that $D$ is '0' and becomes '0' following the condition that $D$ is '1'.

Twisted ring counters are often known as *Johnson code counters* and the code in table 10.5 is a 4-bit Johnson code. The correct logical conditions for a Johnson code sequence can be generated after one complete cycle by energising the input to stage C from a network which develops the logical function $B.(A+C)$, rather than by energising it directly by signal B. You may like to test your skill by verifying this fact.

## 10.12 CHAIN CODE GENERATORS

The binary sequence which is generated by a chain code generator has no apparent logical pattern and, in fact, appears to be a random sequence of binary numbers. The code is generated by a shift register whose input is derived either from a gate or from a network which develops a more or less complicated logical function of the outputs from the register.

A basic form of chain code generator is shown in figure 10.12, and comprises a 4-bit shift register whose input is supplied by the output from a NOT-EQUIVALENT gate. The input is fed in serially at the left-hand end and the output may either be fed out in a serial or in a parallel mode. The code sequence generated by figure 10.12 is given in table 10.6, and we see from the table that a '1' is fed into stage A following the condition that $C \not\equiv D$.

Assuming that the cycle commences with $A = 1$, and all other outputs are equal to zero then, since $C \equiv D$, a '0' is fed into stage A when the first shift pulse is applied. Once more, after the first clock pulse, $C \equiv D$ so that a second '0' is fed into stage A by the second clock pulse. After this pulse condition $C \not\equiv D$ exists and a '1' is returned to the input of the register. This '1' appears at the output of stage A after the third clock pulse has been completed. The code sequence of the counter is completed after fifteen shift pulses have been applied.

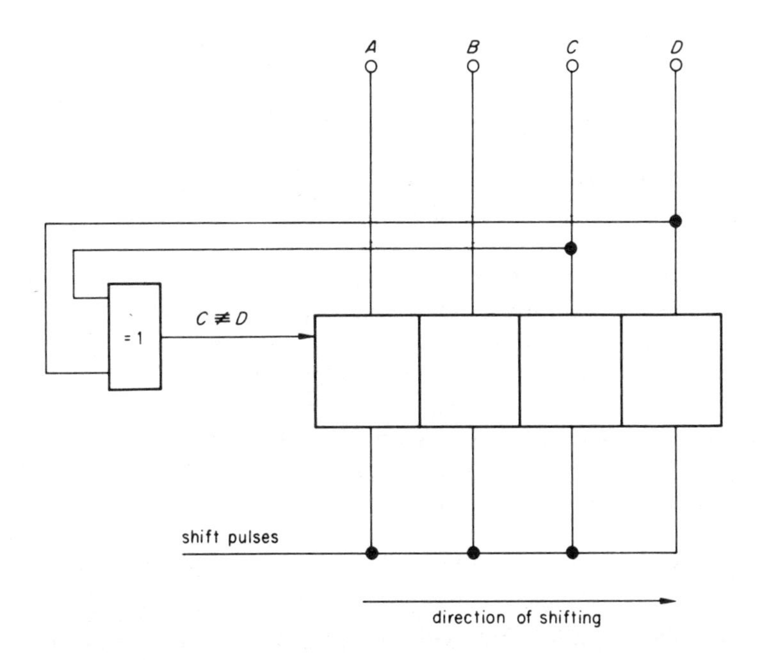


Figure 10.12 A chain-code generator

**Table 10.6**

| *Shift pulse* | *A* | *B* | *C* | *D* |
|---|---|---|---|---|
| initial condition | 1 | 0 | 0 | 0 |
| 1 | 0 | 1 | 0 | 0 |
| 2 | 0 | 0 | 1 | 0 |
| 3 | 1 | 0 | 0 | 1 |
| 4 | 1 | 1 | 0 | 0 |
| 5 | 0 | 1 | 1 | 0 |
| 6 | 1 | 0 | 1 | 1 |
| 7 | 0 | 1 | 0 | 1 |
| 8 | 1 | 0 | 1 | 0 |
| 9 | 1 | 1 | 0 | 1 |
| 10 | 1 | 1 | 1 | 0 |
| 11 | 1 | 1 | 1 | 1 |
| 12 | 0 | 1 | 1 | 1 |
| 13 | 0 | 0 | 1 | 1 |
| 14 | 0 | 0 | 0 | 1 |

This type of shift-register counter utilises the flip–flops to greater effect than counters described hitherto since fifteen of the sixteen possible code combinations are used. The code group missing in table 10.6 is the combination 0000; if the outputs assume the 'all zeros' condition at the instant of switch-on, there is no apparent movement of data since the zeros continuously circulate through the counter. In counters of this type it is necessary to include additional logic to prevent the 'all zeros' condition arising.

Had we used an equivalence gate in the feedback loop of figure 10.12, we should have generated a code sequence of length fifteen which includes the 'all 0s' condition, but which excludes the 'all 1s' state. You may like to verify this fact.

Since the pattern generated by the network is not a truly random pattern, we call it a *pseudo-random binary sequence* (PRBS). Applications of PRBS generators range from illumination control (for example, Christmas tree lighting) to electronic system testing.

The cycle length generated by chain-code generators depends not only on the number of stages over which the feedback is applied, but also on the way in which the signal is derived. The maximum code length for $N$ stages is $(2N - 1)$, which is 7 for three stages, 15 for four stages, 31 for five stages, and so on.

## 10.13 A SERIAL BINARY ADDER

The basis of serial addition was dealt with in chapter 9, and here we describe a complete circuit such as may be used in a calculating machine.

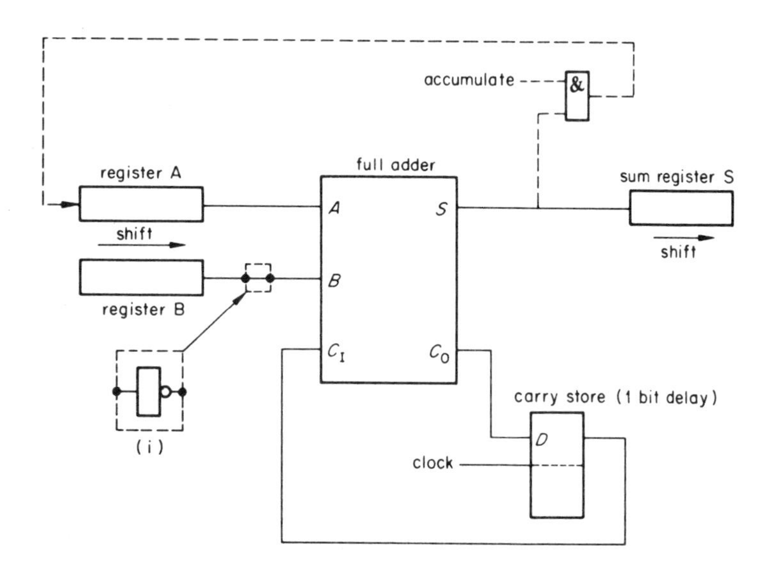


Figure 10.13 A serial binary adder

Registers A and B in figure 10.13 are used to store the two binary words to be added together, and these are shifted serially into the full adder. At the same time, the sum formed by the addition is shifted into the *sum register* S, the carry for the following addition being stored in the carry store, the latter initially storing a '0'. The sum register must be longer than the binary number by one bit in order that any 'overflow' produced during the calculation is not lost.

The number of registers used can be reduced from three to two by including the recirculating circuit shown in broken line in figure 10.13. In this case, when the *accumulate* line is energised by a '1' signal, the SUM output from the adder is shifted into register A. Thus, register A carries out the dual functions of storing one of the numbers to be added and also of storing the final sum. As the number in register A is shifted out of the right-hand end and into the full adder, so the sum is shifted into the left-hand end. When a '0' signal is applied to the accumulate control line and shift pulses are applied, a series of 0s are shifted into register A.

## 10.14 A SERIAL BINARY SUBTRACTOR

We saw in section 9.6 that subtraction can be carried out by the process of adding the complement of one of the numbers. If, in figure 10.13, we include the invertor shown in inset (i), the value accumulated in register S is the binary value of $(A - B)$. In this case the most significant bit stored in register S is the sign bit.

The carry store output must be set to logic '1' before the start of the subtraction process; this ensures that we add the 2's complement of the contents of register B to the contents of register A.

As with the adder, register A can be used for the dual functions of storing the minuend before the numbers are subtracted and the difference after the subtraction.

## PROBLEMS

**10.1** Show how four bistable elements can be used in a pure binary counter. Sketch the waveforms at each output to illustrate the operation of the counter.

**10.2** Compare the merits of asynchronous and synchronous counting systems.

**10.3** The counter in figure 10.14 consists of three cascaded $J$–$K$ flip–flops, each having the $J$ and $K$-inputs connected to logic '1' signals. If $A = B = C = 0$ initially, draw up a table showing how the outputs change for the first eight input pulses.

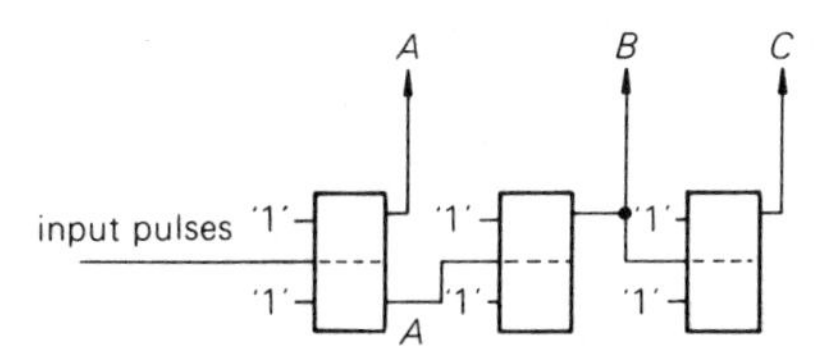


Figure 10.14

**10.4** Design (a) a shift register which shifts 'left' only, (b) a bidirectional shift register.

**10.5** Design a shift register in which data can be parallel loaded into the register.

**10.6** Design a five-stage Johnson code counter in which any spurious code generated at switch-on is eliminated after one complete cycle of events of the counter.

**10.7** Deduce the code sequence generated by a 5-stage chain-code generator with NOT-EQUIVALENT feedback signals from stages 3 and 5.

# 11 Storage Devices and Systems

## 11.1 TYPES OF MEMORY

Many electronic circuits require memory elements to record the state of operation of the circuit at particular instants of time. The most basic form of memory element is the *set–reset flip–flop* (or the S–R flip–flop) (for details see chapter 8). This circuit is widely used not only as a memory element in its own right, but also as the basis of more sophisticated elements known as *master–slave flip–flops* (see section 8.5). The S–R flip–flop is one of a family of memories known as *static memories*, which retain the stored information indefinitely so long as the power supply is maintained; both bipolar and MOS technologies are used in the manufacture of static memories.

Another family of memories, known as *dynamic memories*, use MOS technology, and depend for their operation on the ability of the gate insulation of MOS devices to retain an electrical charge for a relatively long period of time. Ultimately, the charge stored in the gate dielectric decays in value and must periodically be 'refreshed'.

In general, the information stored both in static and in dynamic memories is lost when the power supply fails. Such storage systems are known as *volatile stores*.

A semiconductor memory chip contains an array of cells, usually in the form of a matrix of the type in figure 11.1. Each cell can be independently *addressed* by energising appropriate $X$ and $Y$-address wires. For example, if row wire $B$ and column wire $F$ are energised, then cell $V$ is addressed; by using additional wires (not shown), information can either be *written* into or *read* from the cell. Addressing a particular *location* in the memory in this way is known as *X–Y selection* or *coincident selection*; this allows one *bi*nary digi*t* or *bit* either to be written into or to be read from the selected location. If the organisation associated with the addressing logic is altered, it is possible to access several cells simultaneously. For example, if all three $X$-wires are energised and, if column wire $F$ is also energised, then cells $U$, $V$ and $W$ are simultaneously addressed. This is known as *word selection* or *linear selection*; a binary *word* is a group of binary digits which form the normal unit in which information is stored. In figure 11.1, a word contains three bits. The store shown may be regarded as storing either nine bits of data or three words each of three bits.

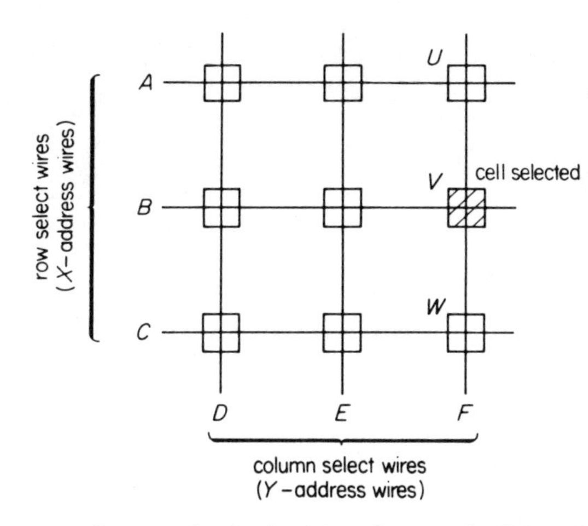


Figure 11.1 One method of addressing an individual cell in a random–access memory

The type of memory in figure 11.1 is known as a *read–write memory* or *random-access memory* (RAM), since the bits (or words) stored may be selected at random. This type of memory is capable of being accessed very rapidly.

Memories known as *read-only memories* (ROM) store data that cannot normally be altered. The data stored in the ROM is frequently specified by the user, and is inserted in the ROM either at the manufacturing stage or, in the case of electrically *programmable ROMs* (PROMs), when it is installed on site. These memories are *non-volatile*, and are used in a number of applications including storing *microprograms* for computer control, and for storing tables (that is, sine, cosine, etc.); also for storing character patterns for use with optoelectronic display devices. The information stored in *reprogrammable ROMs* (RPROMs) can be altered either by electrical or optical methods; the data stored in these devices decay very slowly, and refreshing is required only at very infrequent intervals.

## 11.2 STATIC RAMS

A basic unit or *cell* of the bipolar static RAM is the S–R flip–flop. One form of RAM cell is illustrated in figure 11.2, in which TR1 and TR2 are the active elements in a cross-connected NOR flip–flop. In operation, when the cell is storing data, the $X$ and $Y$-select wires are at a low potential so that the current in the transistor which is *on* flows to both of the 'selection' lines.

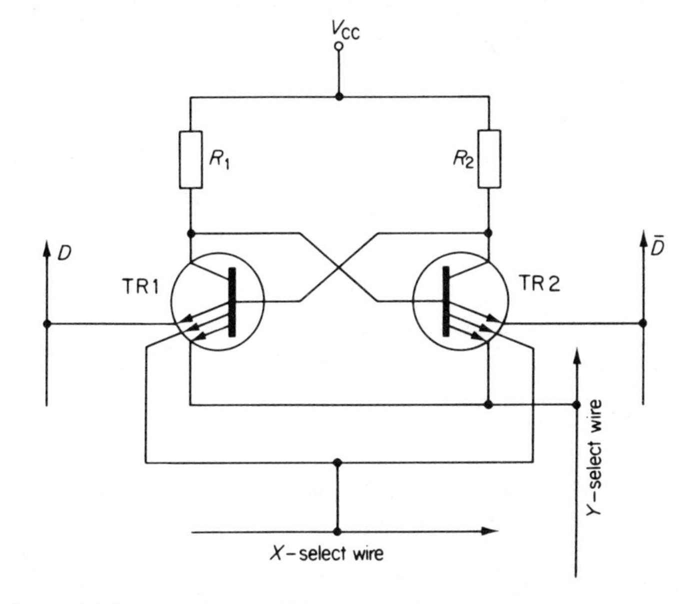


Figure 11.2 One form of bipolar static memory element

To *read* the data stored, both the $X$ and the $Y$-select wires are raised to a positive potential; this causes the emitter current to flow either into the $D$-line or the $\overline{D}$-line (depending on which transistor is *on*). The state of the stored data is therefore indicated by a current pulse in the appropriate data line. To *write* data into the cell, both the $X$ and $Y$-select lines are raised to a positive potential and, if TR1 is to be turned *on*, line $D$ is switched to a low voltage and $\overline{D}$ is raised to a positive potential.

A basic method of organising a nine-bit $X$–$Y$ select static RAM is illustrated in figure 11.3. Each $X$ and $Y$-row is addressed by a single wire, and a common pair of data lines is associated with each cell. By addressing one $X$-line and one $Y$-line, only one cell is

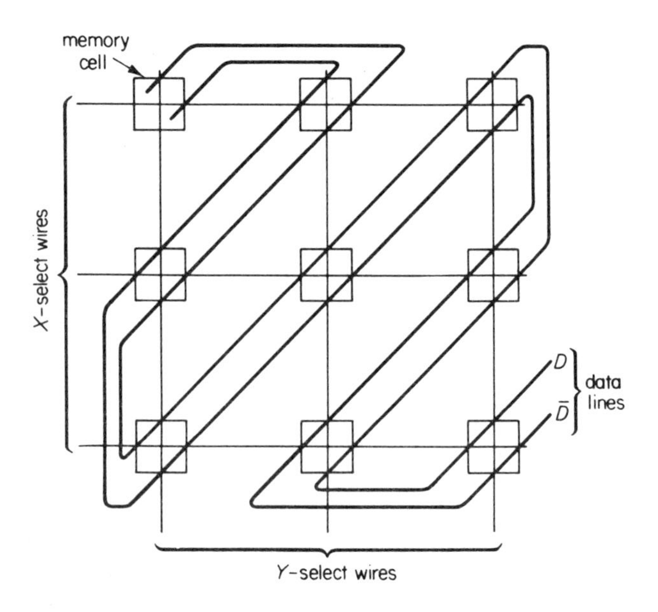


Figure 11.3 A static nine-bit memory matrix

selected; data can then either be read from or written into that cell in the manner described above.

## 11.3 DYNAMIC RAMS

A popular form of dynamic memory cell employing three MOSFETs is shown in figure 11.4. The data is stored in the form of an electrical charge on the gate capacitance, $C$, of transistor TR2. Activating the 'write select' line turns TR1 *on*, and connects capacitor $C$ to the 'write data' line; either a '1' or a '0' may be written into the memory element by connecting an appropriate voltage to this line or, alternatively, the stored data may be 'refreshed'. Since the stored charge in the capacitor ultimately decays, it is necessary to carry out the refreshing operation or recharging operation every few milliseconds.

To read data from the cell, the 'read select' line is energised, which turns TR3 *on* and connects the 'read data' line to TR2. If a

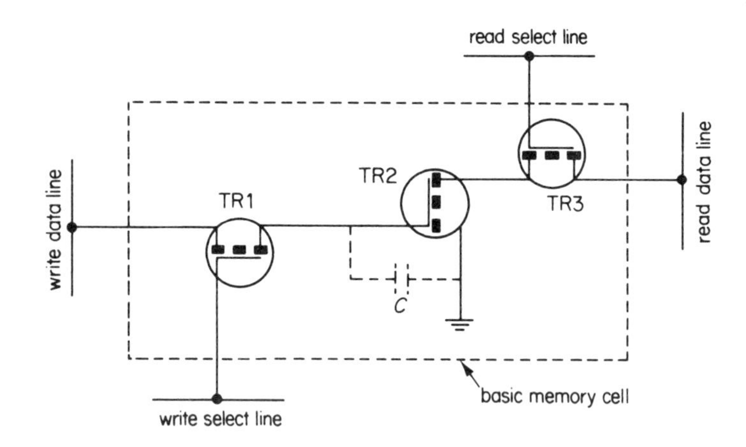


Figure 11.4 A three-transistor dynamic MOS memory cell

logic '1' is stored, TR2 is *on* and current flows in the 'read' line. If a '0' is stored, TR2 is *off* and no current flows in the 'read' line.

One form of organisation of a four-bit dynamic memory matrix is shown in figure 11.5. When the $X_1$ 'read select' line is energised, cells $A$ and $B$ are addressed, and data is read from cell $A$ by monitoring the current in the $Y_1$ 'read data' line. Data is written into cell $A$ by energising the $X_1$ 'write select' line and, simultaneously, applying the data to the $Y_1$ 'write data' line. Data is refreshed by simultaneously reading data from the cell and writing it back in again.

## 11.4 CONTENT ADDRESSABLE MEMORIES (CAMS)

Content addressable memories are designed with the computer programmer in mind, rather than the designer. These memories are addressed by their contents (or a part of their contents), rather than by the physical location within the memory. For example, an employer may wish to know the names of all his employees under 40 years of age, earning over £6000 and with a mathematics A-level qualification. This data can be used to 'address' the CAM.

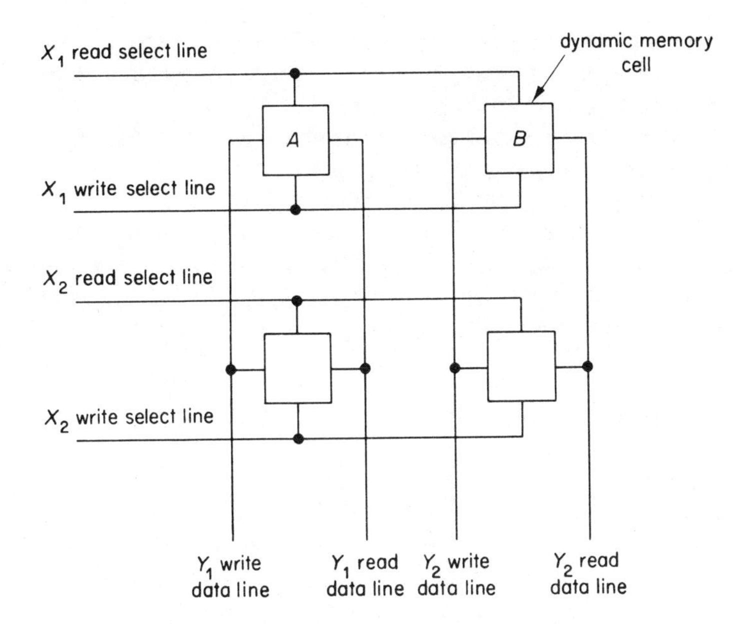


Figure 11.5 Memory organisation for a four-bit dynamic RAM

## 11.5 READ-ONLY MEMORIES (ROMS)

In general, ROMs contain information which is accessed frequently, but is changed only infrequently (or never changed in some cases). The name 'ROM' is perhaps misleading, since the information must be 'written' into them at some time. Some ROMs have data written into them only once during their lifetime, while others have the data changed at infrequent intervals.

There are three broad categories of ROMs, namely *mask programmable*, *electrically programmable* and *reprogrammable*. Versions of the three types are described below.

### Mask Programmable ROMs

The program stored in this type of ROM (which may be specified by the customer) is introduced into the ROM during the manufacture of the IC. Mask programming can be understood from figure 11.6.

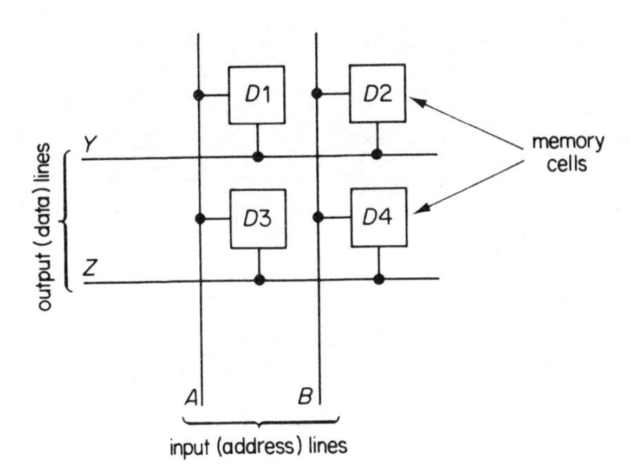


Figure 11.6 The basis of a mask programmable ROM

As outlined earlier, a number of photographic masks are used in the manufacture of monolithic ICs. By altering some of the masks it is possible to cause some of the devices (which are either diodes or transistors) either to be permanently *on* or to be permanently *off*. In this way the memory cell either stores a '0' or a '1'.

The semiconductor memory array in figure 11.6 contains four such cells ($D1$ to $D4$), and if the ROM had been mask programmed so that $D1$, $D2$ and $D3$ are *on*, and $D4$ is *off*, then the application of a logic '1' signal to address line $A$ causes 1s to appear at data lines $Y$ and $Z$. A logic '1' applied to address line $B$ causes a '1' to appear on output line $Y$, and a '0' on line $Z$.

### Electrically Programmable ROMs (PROMs)

A disadvantage of mask programmable ROMs is that the initial manufacturing cost is high unless very large production quantities are involved. To overcome this problem, particularly where small production quantities are involved, electrically programmable ROMs can be used.

A simple PROM is shown in figure 11.7. It is electrically similar to the matrix in figure 11.6, but with each cell in the form of a diode in series with a fusible link. This link may, for example, be in the form of either a thin aluminium link or a polysilicon fuse.

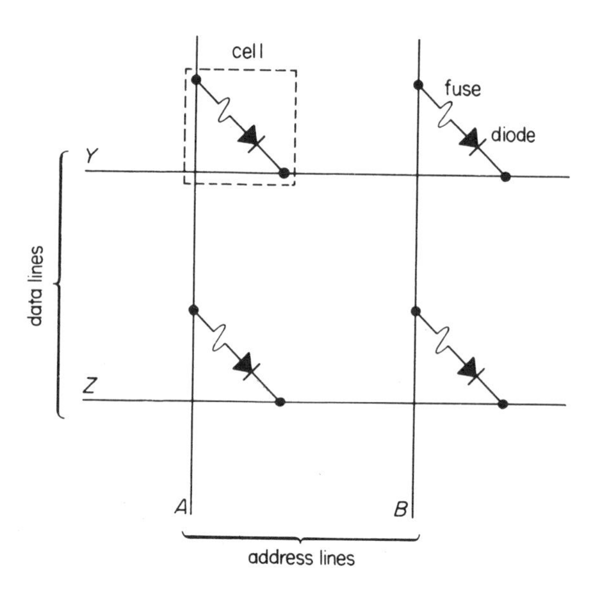


Figure 11.7 One form of PROM

In cases where a diode is to be left in the *on* state, the fuse is left intact. Where it is to be off, the fuse is blown by the application of a current pulse between appropriate address and data lines. This process can be carried out by the manufacturer or by the user by means of suitable electronic apparatus.

### Reprogrammable ROMs (RPROMs)

Reprogrammable memory cells are MOS elements constructed in such a way that the application to the gate region of a voltage above a certain critical value causes charges to be trapped in the insulation below the gate. This produces a conducting inversion layer which links the source and drain of the MOSFET. The amount of charge stored depends not only on the value of the applied voltage, but also on the time interval for which it is applied. Over a period of time the charge slowly leaks away and, for a low value of applied voltage and for a short time-duration, data may be stored for, say, one day. A higher voltage and a longer time-duration cause the charge to be retained for a longer period—a figure quoted by one manufacturer is that a 24 V, 10 ms pulse results in data being retained for more $10^{11}$ read accesses, which may amount to many years in use.

Data may be erased by applying a voltage pulse to the gate, but of reverse polarity to the 'writing' pulse. In other cases, data is erased by exposing the chip to ultraviolet radiation.

Since reprogrammable ROMs are most frequently used as read-only memories, they are also known as *read-mostly memories* (RMM). Another name given to them is *electrically alterable read-only memories* (EAROM).

## 11.6 PROGRAMMABLE LOGIC ARRAY (PLA)

A ROM may be considered to be a combinational logic array, having outputs which are logical functions of the inputs. In this way the ROM can be used to replace one or more logic networks. When used in this way it is known as a programmable logic array. Once the truth table of a particular logic network has been specified, the equivalent logic network can be replaced by a PLA.

## 11.7 STORAGE OF DATA ON A MAGNETIC MEDIUM

A ferromagnetic medium can operate in one of two states, namely magnetised or demagnetised and can therefore be used to store binary information. Moreover, due to the phenomenon of magnetic hysteresis, a magnetic material can be magnetised either in what we arbitrarily describe as the 'forward' direction or in the 'reverse' direction.

Many devices have been used to record data magnetically including ferrite cores, magnetic surface recording (tapes, drums and discs) and plated wire stores. These methods are briefly outlined in the following sections.

## 11.8 FERRITE CORE STORAGE

The type of ferrite core used in storage systems is the ring or

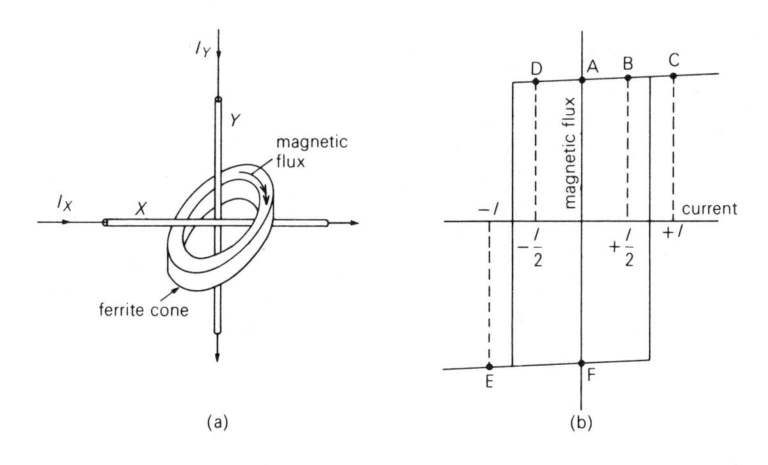


Figure 11.8 A ferrite ring storage element

toroid shown in figure 11.8a. A characteristic of ferrite is that it has a square B–H loop (or $\Phi$–$I$ loop) of the type in figure 11.8b. The outside diameter of the core may typically be 0.4 mm or less.

Suppose for the moment that the core has previously been magnetised so that it operates at point A on the flux–current loop. If we pass a current $I_X$ along the $X$-wire, a flux is established in the core in the direction shown in figure 11.8a. We define the value of the current, $I$, in figure 11.8b as the current which is necessary to magnetise the core. If the value of $I_X$ is $+I/2$, the state of magnetisation of the core is changed from A to B. This is the 'positive' direction of current and is known as a *half-write current* (the reason for this name is described later). When the current is reduced to zero, the operating state A (figure 11.8b) is resumed once more. If $I_X$ is reversed so that it is $-I/2$, the state of magnetisation is changed from A to D. This is the 'negative' direction of current and is known as a *half-read current*. When the current is reduced to zero, the operating state A is once again resumed. Alternatively, we may excite the core by means of current $I_Y$ in wire $Y$, with the same general effect as for current $I_X$. That is, when only one wire carries a 'half-current', the state of magnetisation of the core is virtually unchanged.

However, if we excite both the $X$ and $Y$ wires *simultaneously* with half-currents, the net effect is the same as exciting one wire with current $I$. This results in the operating point on the curve in figure 11.8b moving either from A to C (this current is known as a *full-write current*) when there is little change in the core flux, or from A to E (this current is known as a *full-read current*) when the flux is reversed.

To summarise

(1) a half-current pulse on one wire alone does not result in any significant change in the magnetic state of the core;
(2) a half-current pulse applied to both the $X$ and $Y$-wires simultaneously produces a significant change on the flux-current curve; it may result in a change of the magnetic state of the core (depending on the original state of magnetisation and on the direction of the current pulses).

## 11.9 READING FROM AND WRITING INTO A FERRITE CORE

In order to interrogate or to *read* the magnetic state of the core, we need to introduce an additional wire known as the *sense wire* (see figure 11.9a). The function of this wire is as follows. When reading data from a core store we use a process known as *destructive readout* (DRO). In this process, the data stored in a particular location (that is in a particular ferrite ring) is forced to the logic '0' state by the application of half-read current pulses on the $X$ and $Y$-wires simultaneously. If the core originally stored logic '0', then this results in only a small change in the magnetic flux in the core. The net result in this case is that only a small voltage is induced in the sense wire which passes through the core (see the waveforms for *read '0'* in figure 11.9b).

If, after reading the data in the store location, we wish to write a '1' into the core (see waveforms for *write '1'* in figure 11.9b), it is necessary to reverse the direction of the half-current pulses in the $X$ and $Y$-wires to give the effect of a full-write current. The net result is that the flux in the core is reversed in the manner outlined earlier. This causes the operating point to change from point F in figure 11.8a to point A.

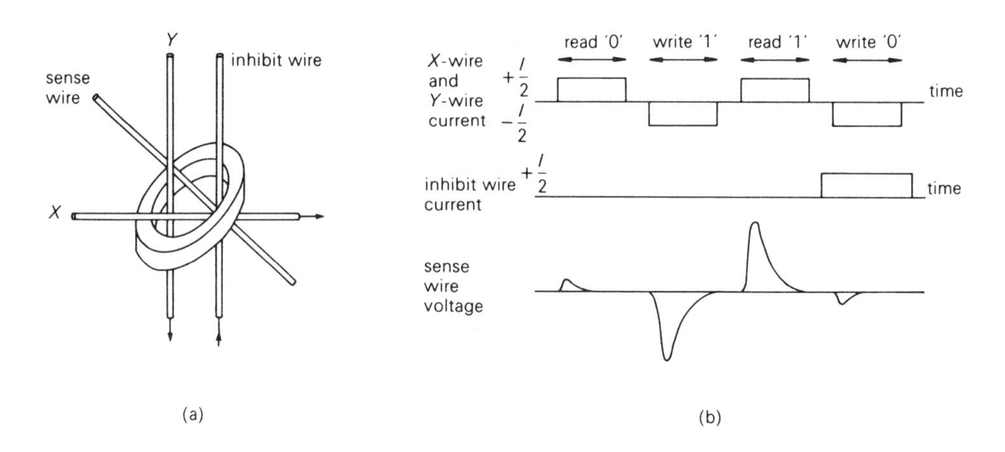


Figure 11.9 Reading data from and writing data into a ferrite ring

In order to read the '1' stored in the core (see figure 11.9b), we apply half-read current pulses on the $X$ and $Y$-wires which effectively 'write' a '0' into the core. This process changes the operating point on the flux-current curve from A to F and a large voltage pulse is induced in the sense wire.

Let us now consider how we write a '0' into the core. At the end of each read cycle the core contains a logic '0'; in order to prevent a '1' being written into the core during the next write cycle, we must cancel out the effect of one of the half-write pulses. The net result is that only one half-write current pulse is applied to the core wires and, as we have seen earlier, this does not alter the magnetic state of the core. Cancelling one of the half-currents is achieved by means of a current pulse in another wire known as the *inhibit wire*, which passes through the core (see figure 11.9a) and is parallel with the $Y$-wire. In order to write a '0' into the core, a half-current pulse is passed along the inhibit wire during the 'write' period cycle (see the waveforms for *write '0'* in figure 11.9b) so that it cancels out the magnetic field produced by $I_Y$.

Since destructive read-out is used, it may be necessary to re-write the data back into the store after it has been 'destroyed' in the reading cycle. The complete cycle of reading and writing is known as the *read–write cycle*.

## 11.10 ADDRESSING A LOCATION IN A CORE STORE

One of the most popular methods of organising data storage in a core store is the *coincident-current* method or 3D method illustrated in figure 11.10. The store is organised in storage *planes*, the planes being stacked one above the other as shown. In the case considered, each plane contains a 4 × 4 matrix of ferrite rings, giving 16 locations per plane. In order to address each storage location, four $X$-wires and four $Y$-wires run through each plane. The binary word which will be addressed in the manner outlined below is located in cores $B_1$, $B_2$, $B_3$ and $B_4$. To simplify the diagram, the sense and inhibit wires are omitted; there is one sense wire *per plane* and one inhibit wire *per plane*, both wires linking with every ferrite ring in one plane.

To simultaneously address bits $B_1$ to $B_4$, half-current pulses $I_X$ and $I_Y$ are applied to the respective $X$ and $Y$-wires. Data is then written into or read from the storage locations $B_1$ to $B_4$ in the manner outlined in section 11.9. Note that in the upper plane, cores $X_1$, $X_2$ and $X_3$ are excited by the half-current $I_X$, which produces no change in the magnetic state of those cores. Similarly, half-current $I_Y$ does not affect the magnetic state of cores $Y_1$, $Y_2$ and $Y_3$.

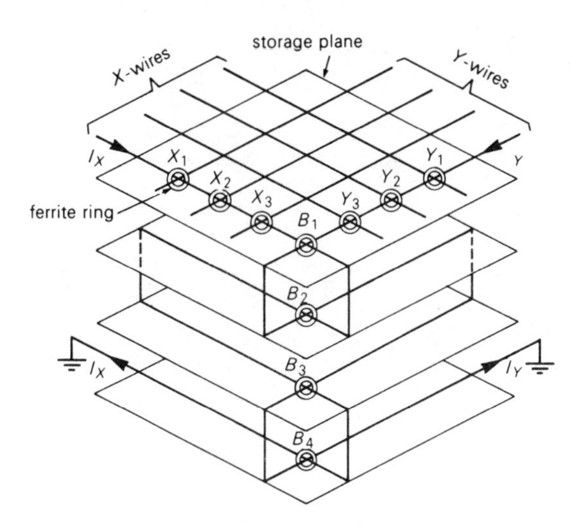


Figure 11.10 A coincident-current (3D) storage stack

## 11.11 MAGNETIC SURFACE RECORDING

Binary data can be stored on magnetic medium deposited on the surface of tapes, discs, drums and wires. Magnetic discs fall into two categories, namely rigid discs and floppy discs; in rigid discs the magnetic medium is deposited on a rigid backing whereas it is deposited on a flexible backing in floppy discs. In a plated-wire memory the magnetic material is deposited on a wire which is part of the read–write circuit.

## 11.12 METHODS OF RECORDING DATA ON A MAGNETIC SURFACE

Many methods have been adopted to record binary data on magnetic tapes, discs and drums, and we will limit the discussion here to two techniques, namely one method of return-to-zero recording and one method of non-return-to-zero recording.

### Return-to-zero (RZ) Recording

In RZ methods of recording, the magnetisation level returns to zero after recording each bit. Typical flux waveforms are shown in figure 11.11a. RZ methods of recording data are not very efficient in terms of the length of magnetic surface required to store the data. The reason is that two flux reversals (that is, an increase in flux and a decrease in flux) are required to record each binary digit. However, it has the advantage over some other methods of recording in that it is *self-clocking*, that is the recorded pulses can be used to generate its own sequence of clock pulses. Arising from this, the speed at which the magnetic surface moves under the read–write head is relatively unimportant.

One limit to the data packing density on the magnetic surface is the rate at which the magnetic flux can be increased to its maximum value and then reduced to zero again, and yet produce a clearly identifiable pulse in the reading head.

### Non-return-to-zero (NRZ) Recording

NRZ methods are so named because the writing-head current does not return to zero after recording each bit and the magnetic medium is saturated either in one direction or in the reverse direction.

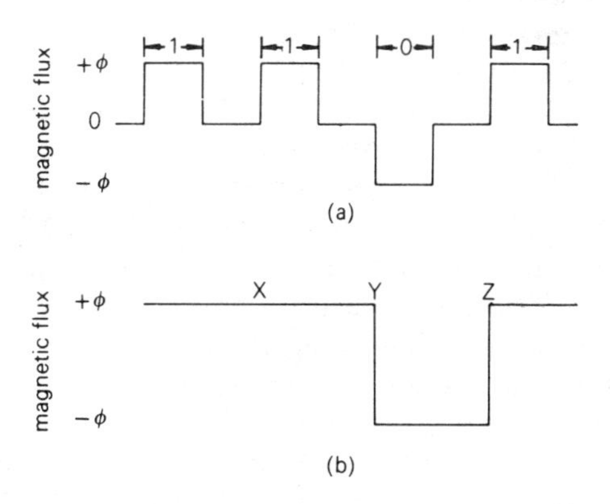


Figure 11.11 (a) One method of RZ recording and (b) one method of NRZ recording

In the NRZ method illustrated in figure 11.11b, the flux changes only when the bit pattern changes. For example, at point X the bit pattern remains the same so that there is no change in the recorded flux, whereas at point Y it changes from '1' to '0', and at Z it changes from '0' to '1'. Thus each pulse from the reading head merely indicates that the bit pattern has changed; the logic system must therefore 'remember' whether a '0' or a '1' was recorded previously.

The most important feature of the NRZ method when compared with the RZ method is that a given pattern of bits can be recorded in half the length of recording medium which is required for the RZ method. This arises from the fact that the number of flux reversals in the NRZ method is at least one-half that of the RZ method, so that it can be recorded at twice the speed of the RZ method. A disadvantage of the NRZ method is that the recorded pattern is not self-clocking.

## 11.13 MAGNETIC BUBBLE DEVICE (MBD) MEMORIES

In ferromagnetic materials, the atoms are associated in groups known as *domains*, each domain appearing as though it were a small permanent magnet. In a demagnetised material, the domains point in random directions, the 'randomness' being such that the net magnetic field of all the domains cancels out.

When the magnetic field applied to a ferromagnetic material is small, the domains begin to align with the magnetic field and just before the material saturates, the remaining domains which have not aligned with the field snap into stable cylindrical 'bubbles' (see figure 11.12). A further increase in field strength causes a reduction in the diameter of the bubbles and, finally, at a critical field strength the bubbles collapse leaving the material completely magnetised in the direction of the applied field. The latter process is sometimes used as a means of annihilating magnetic bubbles. Under the conditions which allow magnetic bubbles to exist, binary data can be stored simply as the presence or absence of a bubble.

The garnet family of materials is used as the magnetic medium in which the bubbles are produced, and a patterned layer of magnetic material such as Permalloy (NiFe) is deposited on the surface of the garnet to provide a path along which the bubbles are shifted or

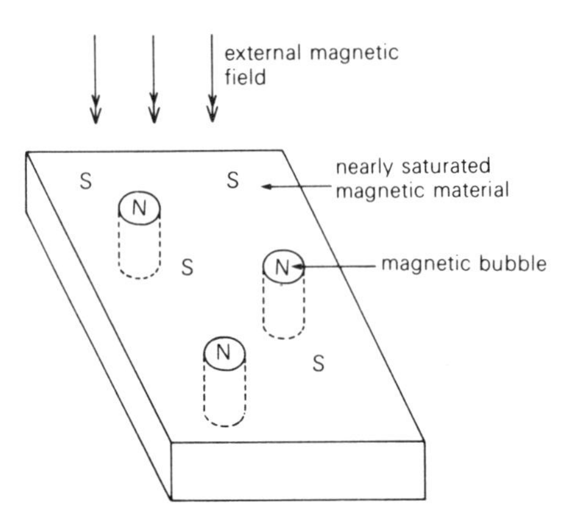


Figure 11.12 Bubble domains

guided. A plan view of a small section of a bubble memory (or shift register) is shown in figure 11.13, the T-bar and I-bar Permalloy sections being deposited on the surface of the garnet in which the bubble is situated. The process of bubble generation is described later and for the moment we will concentrate on the *propagation* of the bubble.

The bubble memory is situated in a magnetic field which rotates in the plane of the magnetic surface. The rotating field is produced in much the same manner as that in an induction motor. At the instant when the rotating field acts in the direction shown in figure 11.13a, the magnetic polarity of end A of the upper T-bar attracts the bubble. When the magnetic field has rotated through an angle of 90° (to the direction shown in figure 11.13b), the main polar attraction causes the bubble to move to point B which is at the end of the horizontal part of the T-bar. A further 90° rotation of the magnetic field (see figure 11.13c) causes the bubble to be attracted to end C of the T-bar. Yet another 90° rotation of the field (figure 11.13d) results in the bubble transferring to end D of the centre I-bar. Another 90° rotation of the magnetic field causes the bubble to transfer to end E of the lower T-bar. Other shapes than T and I-bars are used in some practical systems.

An advantage of bubble memories over their semiconductor counterparts is that they are non-volatile. Moreover, power can be

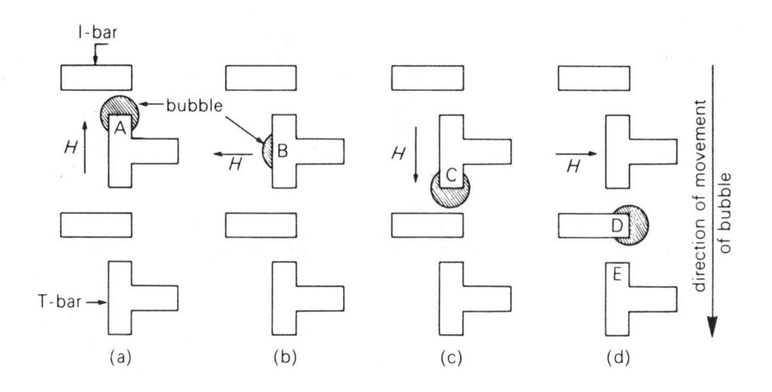


Figure 11.13 Bubble propagation

conserved, if the bubbles do not need to be moved, simply by switching off the rotating field.

In addition to the bubble-propagation mechanism described above, the basic requirements of a bubble memory include bubble generation, bubble detection and bubble annihilation, which are described below.

*Bubble generation* is the process of 'writing' information into the memory. One method of doing this is to pass a pulse of current around a hairpin-shaped conducting loop, the bubble being formed in the end of the loop. One method of *bubble detection* is the use of a material exhibiting *magnetoresistance*, which is a property of materials whose resistance is altered by a magnetic field. Permalloy is one such material. When a magnetic bubble arrives in the detector, it is 'stretched' or is distorted to cause a distinct change in the magnetoresistance of the Permalloy. Unwanted bubbles can be *annihilated*, a simple method being to raise the local magnetic field strength above the critical value to cause bubble collapse. However, other techniques can also be used.

## 11.14 CHARGE-COUPLED DEVICE (CCD) MEMORIES

A charge-coupled device (CCD) is a multigate MOS device capable of transferring electrical charge from its source electrode to its drain electrode in the form of a series of charge 'packages'. Charge-coupled devices have properties which are particularly useful in the fields of memory devices, in analogue delay lines and in solid-state optical imaging.

The principle of charge transfer can be understood from figure 11.14. Figure 11.14a shows a basic *p*-channel MOS CCD (*n*-channel devices are also manufactured). This device has an *n*-type substrate, and the application of a negative potential to the electrode on the surface of the CCD repels the mobile negative majority charge carriers away from the underside of the oxide. This results in the formation of a depletion region below the electrode. As the electrode potential is increased, the depletion region extends further into the substrate; at the same time, the negative potential on the gate attracts minority charge carriers (holes) to the depletion region until, at the threshold voltage, these minority carriers form a conducting channel at the oxide–semiconductor interface.

For the purpose of explaining the operation of the device, it is convenient to regard the depletion region as being a *potential well* (see figure 11.14a). Also, even though the 'holes' in the inversion layer are located immediately below the oxide, it is convenient to

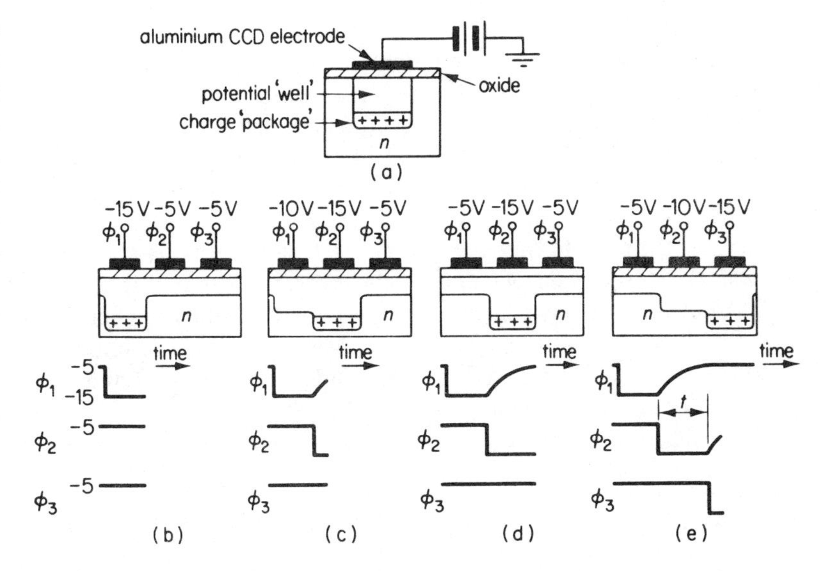


Figure 11.14 Operating principle of charge-coupled devices

think of them 'filling' the lower part of the potential well. Thus, the charge 'package' is constrained in the well.

In the basic CCD structure (diagrams (b) to (e) of figure 11.14), three electrodes are required to complete the transfer of one charge packet. In consequence the three electrodes are referred to as one *element* of the CCD. The gap width between the electrodes is kept as small as possible to give a reasonable value of *charge transfer efficiency*, $\eta$ (which usually has a value in the range 99.9 to 99.99 per cent), between the elements; the value of the voltage applied to the electrodes is usually greater than the threshold voltage of the device. The mechanism of charge transfer is described in the following.

The drive lines supplying the electrodes are energised by a three-phase supply, the basic voltage waveforms applied to $\phi_1$, $\phi_2$ and $\phi_3$ are as shown in the figure. Figure 11.14 illustrates the instant of time when the potential applied to $\phi_1$ is greater than that applied to either $\phi_2$ or $\phi_3$, so that the potential well is deepest under $\phi_1$. In consequence the charge package is located under $\phi_1$. A short time later (figure 11.14c), the potential applied to $\phi_2$ is increased, and that applied to $\phi_1$ is reduced. This results in the charge packet transferring once more to the 'deepest' part of the potential well, this time under $\phi_2$. When the potential applied to $\phi_1$ has decayed to its smallest value (figure 11.14d), the potential well has transferred completely to $\phi_2$. A little time later, the potential applied to $\phi_3$ is increased (figure 11.14e), and that applied to $\phi_2$ is reduced, and the charge packet is once more moved to the right. Since the charge cannot be transferred in zero time, it is necessary to allow the potential applied to the electrodes to decay slowly; the time, $t$, of this overlap is known as the *overlap time* (figure 11.14e).

A basic form of a two-element *three-phase CCD* is shown in figure 11.15. It consists essentially of two elements of the type described above, together with a means of injecting charge packets and a means of collecting them.

A CCD memory can be organised along the lines of a recirculating shift register, in which the data is retained by recirculating it from the end of the register back to the beginning.

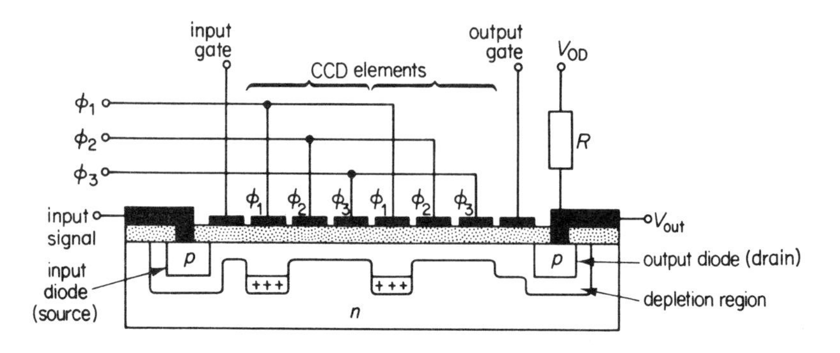


Figure 11.15 One form of three-phase CCD structure

## PROBLEMS

**11.1** Describe the operation of a static memory matrix suitable for implementation in MSI form.

**11.2** Discuss the features of a dynamic RAM with reference to its use in a digital system.

**11.3** Describe the operation of a data storage system using ferrite ring magnetic cores.

**11.4** Discuss possible sources of electronic noise in a core store system. Suggest methods of overcoming the effects of the noise.

**11.5** Devise a logic system for reproducing the data stored on magnetic tape using (a) RZ recording, (b) NRZ recording.

# Solutions to Selected Problems

## Chapter 1

**1.1** (a) 3; (b) 8; (c) 12; (d) 16

## Chapter 2

**2.5** $K = A . B . C$
**2.6** (a) $F = A . B$; (b) $F = A + B$

## Chapter 3

**3.1** (a) $A + B$; (b) $A . B$; (c) $\overline{A + B}$; (d) $\overline{A . B}$

## Chapter 4

**4.2** (a) 2350 Ω-m; (b) 1720 Ω-m

## Chapter 5

**5.3** $R_L = 1\ \text{k}\Omega$; $R_B = 22\ \text{k}\Omega$
**5.8** $A . B . C . D$

## Chapter 6

**6.8** $\overline{A} . B$
**6.9** $B . C$

## Chapter 7

**7.1** $f = B + A . \overline{C} + \overline{A} . C$
**7.4** $f = A . B + A . C + B . C$
**7.6** (a) $A = 0$, $B = 0$; $A = 1$, $B = 1$; (b) $A = 1$, $B = 0$; $A = 0$, $B = 1$
**7.7** $A = 1$, $B = 0$, $f = 0$

## Chapter 9

**9.1** 1011100.111
**9.2** 27.875
**9.3** (a) $101011001000_2$; (b) $5310_8$; (c) $2760_{10}$
**9.5** (1)0011.11; (1)00.01; (0)110000011; (1)0011100
**9.7** 0011 0111.0101; 1000 1001 0111; 0010 0000 1000 0111.0101

## Chapter 10

**10.3** 000, 110, 010, 101, 001, 111, 011, 100, 000